*Grover E. Murray*
Studies in the American Southwest

ALSO IN THE GROVER E. MURRAY STUDIES IN THE AMERICAN SOUTHWEST

*Brujerías: Stories of Witchcraft and the Supernatural in the American Southwest and Beyond*
Nasario García

*Cacti of Texas: A Field Guide*
A. Michael Powell, James F. Weedin, and Shirley A. Powell

*Cacti of the Trans-Pecos and Adjacent Areas*
A. Michael Powell and James F. Weedin

*Deep Time and the Texas High Plains: History and Geology*
Paul H. Carlson

*From Texas to San Diego in 1851: The Overland Journal of Dr. S. W. Woodhouse, Surgeon-Naturalist of the Sitgreaves Expedition*
Edited by Andrew Wallace and Richard H. Hevly

*Javelinas*
Jane Manaster

*Little Big Bend: Common, Uncommon, and Rare Plants of Big Bend National Park*
Roy Morey

*Picturing a Different West: Vision, Illustration, and the Tradition of Austin and Cather*
Janis P. Stout

*Texas Quilts and Quilters: A Lone Star Legacy*
Marcia Kaylakie with Janice Whittington

# PECANS

# PECANS

## The Story in a Nutshell



Jane Manaster

Texas Tech University Press

First cloth edition © 1994 as *The Pecan Tree* by the University of Texas Press

This book is typeset in ITC Galliard. The paper used in this book meets the minimum requirements of ANSI/NISO Z39.48-1992 (R1997). ♾

Library of Congress Cataloging-in-Publication Data

Manaster, Jane.
[Pecan tree]
Pecans : the story in a nutshell / Jane Manaster. — 1st paperback ed.
p. cm. — (Grover E. Murray studies in the American Southwest)
Previously published as: The pecan tree. 1994.
Summary: "Brief exploration of the natural history, cultivation, and uses of pecans, both nut and tree. Also describes life cycle, predators and diseases, species development, and commercial expansion. Includes recipes for pralines, pecan logs, roast and candied pecans, and pecan pie"—Provided by publisher.
ISBN 978-0-89672-640-6 (pbk. : alk. paper) 1. Pecan. 2. Pecan—Texas. I. Title. II. Series.
SB401.P4M36 2008
634'.52—dc22

2008030498

*Printed in the United States of America*

08 09 10 11 12 13 14 15 16 / 9 8 7 6 5 4 3 2 1

Texas Tech University Press
Box 41037 | Lubbock, Texas 79409-1037 USA
800.832.4042 | ttup@ttu.edu | www.ttup.ttu.edu

*Frontispiece:* from *Famous Paper Shell Pecan Commercially Grown* (advertising brochure, San Antonio, 1913[?])

*To Guy, with love*

# CONTENTS

*Acknowledgments* xi

Introduction and Range 1

Botanical Niche 7

History 13

Cultivation and Improvement 23

Texas: A Case Study 40

Orchards 52

Animal Predators 60

The Pecan Industry 67

Nutrition 77

Recipes 82

Conclusion 96

*Resources* 101

# ACKNOWLEDGMENTS

I would like to thank Tommy Thompson of the USDA Pecan Research Station in Sommerville, Texas. He explained many aspects of pecan culture with unfailing patience and clarity, after seeing my horrified expression when he first described the pecan tree as a "monoecious, diploid species with a system of heterodichogamy." The book came about thanks to the encouragement of Robin Doughty, a fellow Brit who also became enamored of Texas and the American Southwest.

# PECANS

# INTRODUCTION AND RANGE

People are hitting at trees with sticks. Their hair awry and a compulsive gleam in their eyes, they are hurling sticks in the air, dropping on hands and knees, stuffing their pockets and plastic sacks, then starting the whole procedure again. What esoteric religious ritual can this be? Or are they simply enjoying a primitive exercise routine, intoxicated by the first breath of cool air after the stifling summer heat? The behavior seems most intense along the rivers and streams of the southeastern states, where majestic trees tower in dense groves.

This is indeed a ritual, a seasonal rite, though hardly a religious one. They are gathering pecans as the settlers did, and the first Americans, ancestors in spirit who scrabbled for the crop each autumn among the falling leaves. Back then, the harvest was free for the taking. Now the wily owners of groves and orchards invite visitors who are eager to get back to nature to pick their own pecans and save pennies on the price at the roadside stands, a dollar or more on ready-packed, slightly dusty ones at the supermarkets.

The pecan is among the few native plant species of North America to be developed into a sizable agricultural crop. It is a very profitable business in several southern states, reaching from Georgia in the east and as far west as Arizona. Today most of the pecans in the United States, whether bought in the shell, consumed as ingredients in bakery goods and ice cream, or relished in pralines and candy bars, are cultivars, improved varieties produced by years of patient research. But there remain, firmly rooted in the sandy loam of the river bottoms, uncounted millions of wild seedling trees whose nuts, even if less predictable than their cultivated cousins, still produce bountiful harvests. Long before the genetic restructuring brought about during the last century by grafting extended the pecan belt way beyond its native limits, early Americans included pecans in their diet, especially in the fall. They planted trees to provide cool shade in the summer and used the hard wood for furniture, farm implements, and other needs that arose.

The word *pecan*, meaning "a nut too hard to crack by hand," was used by the Algonquians who ranged from the northern Mississippi to the mouth of the Ohio River, into Kentucky and northern Tennessee, covering the eastern domain of the pecan's natural habitat. Among other eastern woodland tribes the name was singled out for the pecan alone. This name did not reach European ears until the eighteenth century, though the nuts were described in reports and journals from the time of early Spanish exploration along the Texas Gulf Coast. In his diary Cabeza de Vaca, who was in the territory of present-day Texas from 1529 to 1535, compared what was almost irrefutably the pecan to the Spanish walnut, and his colleague, Lope de Oviedo, referred to the "river of nuts"—most probably the Guadalupe River. Spanish explorers of this period, among them Hernando de Soto, who discovered the Mississippi River, identified the pecans in 1541 as either *nueces*, meaning walnuts, or *nogales*, the Spanish word for walnut trees. They noticed that the nuts were an important part of the Indian diet each fall, when the nuts burst from their husks and scattered along the river bottoms.

Early in the eighteenth century the French introduced their own version of the word from which we derive *pecan*. Once thought to have originated as a Creole term, it was in fact adopted through contact with the Algonquians and neighboring tribes. André Pénicaut came to the Lower Mississippi wilderness as a ship's carpenter with Pierre Le Moyne d'Iberville's expedition. Later in life he recalled the first years of European settlement in the province between 1694 and 1722. He described the time he spent in the small Indian village of Natchez: "The natives have three kinds of walnut tree; some whose nuts are as big as the fist from which bread for their soup is made; the best ones, however, are scarcely bigger than the thumb and are called pecan." He spelled the nut *pacane.* So did Xavier Charlevoix, the French missionary and traveler who sailed down the Mississippi and arrived in New Orleans in 1722. Charlevoix's journal entry dated October 20, 1721, explained: "The pecan is a nut of the length and of the form of an acorn. There are some whose shell is very thin; some others have it harder and thicker, this is to the detriment of the fruit; they are even somewhat smaller. All have a very fine and delicate taste. The tree which bears them grows very high; its bark, the odor and form of its leaves have appeared to me similar to those of the European walnut trees." Later Spanish explorers, such as Antonio de Ulloa, began to use the term *pacanos*, and the French word was modified to *pecane.* Webster's Ninth New Collegiate Dictionary lists pecan as "\pi-'kän, - 'kan; 'pe-,kan\ n [of Algonquian origin; akin to Ojibwa *pagân*, a hard-shelled nut] (1773)."

Along the Mississippi Valley from Kaskaskia in southern Illinois to the Gulf of Mexico, the French and Spanish accepted and used the Indian word. In the East, however, the word was unknown until some years later. Colonists there opted for the popular names such as "Illinois nut" and "Illinois hickory." In his diary entry of March 11, 1775, President Washington refers to the "Mississippi nut," but by 1794 it had become the "poccon."

Although the spelling was confirmed without much to-do, the pronunciation has always been whimsical. Algonquian dialects, or perhaps the less finely tuned European ear, variously enunciated the word "päkô'n," "pûkâ'n," "pägâ'n," "pûgâ'n," "päkánn." In the West and South today it is pronounced "pekáhn," and farther east quite firmly "pécan." Those preferring one pronunciation have feigned an inability to understand the other, but the rivalry tends to be jocular.

The United States Geological Survey lists numerous features on the landscape that include "pecan" in their names, especially through the tree's natural range. Texas alone has more than seventy streams named for the pecan. In Texas and elsewhere "pecan" can be found naming not only streams but also creeks, branches, springs, bayous, and other subsidiary waterways. There are many "pecan" schools, churches, parks, and cemeteries, a swamp in Louisiana, islands in Illinois, Louisiana, and Texas, gardens in Virginia, a Tennessee picnic and camp ground, a Louisiana oilfield, and the historical Pecan Post Office building in Mississippi.

A high honor accorded the pecan was its selection as the state tree of Texas, as a result of the dying wish of James Hogg, the state's first native-born governor. On March 1, 1906, Governor Hogg lay on his deathbed. He told the friends surrounding him, "I want no monument of stone or marble, but plant at my head a pecan tree and at my feet an old fashioned walnut . . . and when these trees shall bear, let the pecans and walnuts be given out among the plain people of Texas, so that they may plant them and make Texas a land of trees." Within hours of this proclamation he died, on March 2, Texas Independence Day. Galvanized by this omen, the friends resolved to get in touch with nut growers around the state and organize a meeting in Austin. They decided to plant Russell and Stuart variety pecan trees beside his grave in the Austin city cemetery. Their meeting led to the organization of the Texas Nut Growers' Association at the end of May that same year. In 1919 the Thirty-sixth Texas Legislature passed a resolu-

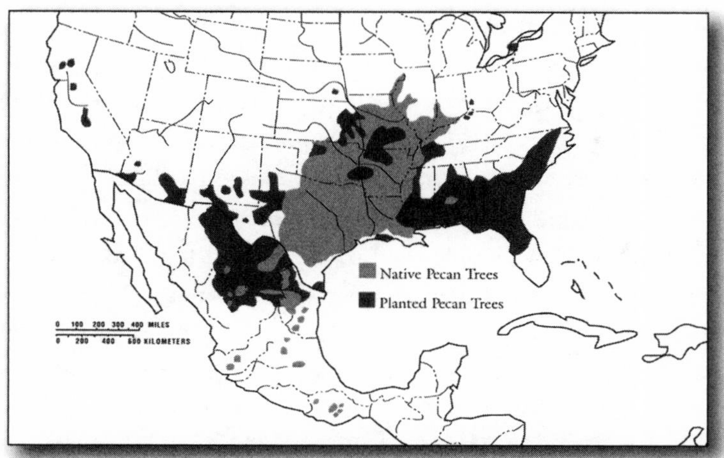


Distribution of native and planted pecan trees in North America. *Map courtesy of L. J. Grauke and T. E. Thompson, USDA-ARS Pecan Breeding and Genetics.*

tion declaring the pecan the state tree of Texas, and this was reaffirmed, for good measure, by the Fortieth Legislature in 1927. In 1969 the Texas Pecan Growers' Association, in cooperation with the late governor's daughter, Ima Hogg, planted a Choctaw pecan tree at the grave site.

The western boundary of the pecan's native range is in Schleicher County, Texas, on U.S. Highway 277 around the small West Texas community of Eldorado, twenty miles north of Sonora on Interstate 10, and some forty-three miles south of San Angelo. The limits of the natural range have been gathered from official and personal reports dating back to sixteenth-century explorers and from later scientific revelations.

The tree's natural range does not extend north of a latitude of about 42°51' on the Mississippi. Generally the northern range is placed at Davenport, Iowa, but with no nuts produced from the native trees north of Clinton, Iowa. From here trees are found growing southward in the Mississippi Valley, and both east and west along its tributaries. They are reported up the Ohio to

Cincinnati, up the Wabash to Terre Haute, Indiana, and, some claim, up the Tennessee as far as Chattanooga. The range extends well up the Arkansas into Oklahoma, and along the Red River, which forms the Texas-Oklahoma boundary before meandering into southern Arkansas and northern Louisiana. Pecan trees are common in southern Oklahoma along the Canadian and Washita rivers, and in Texas beside the Brazos, Colorado, Guadalupe, and other streams. In smaller numbers the pecan borders the Rio Grande, crossing to northern Mexico. While some believe Oaxaca, at latitude 16°30', to be the southern limit of the native range, this is disputed. Some believe the isolated packets of pecans in southern Mexican states date back to the geological Tertiary period or to the time when glaciation chilled northern areas during the Pleistocene epoch. Others consider that people played a part in their diffusion far more recently. Researchers have brought pecans to the National Clonal Germplasm Repository in Texas from eight Mexican states, not so much to unravel their origins as to view the possibility of including some of their distinctive characteristics in breeding programs.

Since cultivation of the pecan began formally toward the end of the nineteenth century, orchards have been planted as far east as the Atlantic seaboard and west into Arizona and California, far from the floodplains that nurture the trees in their natural habitat. There have been attempts, some measurably successful, to grow pecans commercially in South Africa, Israel, Australia, Mexico, and several South American countries. In Japan, several horticulturalists have experimented with pecans in their gardens, and trees have been planted at the Hiroshima memorial. Pecans have even reached the moon, taken there by astronauts on two of the Apollo missions.

# BOTANICAL NICHE

The pecan traces its family history to geological beginnings that long predate North America's present-day morphology. At one time the rim of the Gulf of Mexico reached up to the mouth of the Ohio River, then dipped in a curve that left the northern edge of the Mississippi and Pecos valleys sharing the same Gulf shoreline to the northwest of Texas. Pecan trees growing along the river bottoms from the eastern banks of the Mississippi to the rivers of central Texas thrust their deeply ridged trunks a hundred feet up and more amid the oaks and cottonwoods. A canopy of branches, among the last to leaf each spring, spread a great dome of dark yellowish-green leaflets. The first pecans likely traveled southward along the Ohio and Mississippi rivers and were washed by ocean currents along the Gulf Coast to reach the limits of their native range. Only their presence in western Alabama—perhaps explained by a shift in these earlier ocean currents—challenges this pattern of diffusion. Animals and, much later, people recognized the pleasurable addition of pecan nuts to their diet and reinforced the native range that extends along the floodplains of the Mississippi, Ohio, Missouri, and Red rivers and their tributaries, and the largest rivers of East Texas and northeastern Mexico.

Early explorers and botanists admiring one of the largest trees east of the Rocky Mountains were uncertain where it fitted in scientific taxa. It was confirmed only when their reports were debated that the pecan belongs to the hickory genus. The seventeen members of the tribe Hicoreae belong to the walnut family, Juglandaceae. Most types of hickory are found in North America, but one, *Hicoria cathayensis*, grows just in southern China, probably diffused across Beringia and then taken east by way of Siberia. Juglandaceae is a grand old family, boasting pollen grains that date from the earliest trace of flowering plants in the Cretaceous period 135 million years ago. Hicoreae evolved through intermediate species about 70 million years ago, leaving traces in Eurasia and North America during the Eocene epoch. Juglandaceae genera from the Paleocene and Eocene epochs have been found in North America, with fossilized Hicoreae fruits identified as far apart as Germany and the state of Colorado, where they have been spotted in lower Oligocene strata, and also in China, in Miocene strata, dating back 20 million years. Today the native range of the pecan reaches west only to central Texas.

The first botanical name given to the pecan, *Juglans pecan*, appeared in Humphrey Marshall's *Arbustrum Americanum* published in 1785, but by 1787 a certain Captain Wagenheim who saw the nuts, popularly known as "Illinois nuts," cultivated by the nurseryman William Prince, referred to them as *Juglans illinoensis*. In 1819 Thomas Nuttall, an English botanist, separated the hickories from the walnuts and placed them in a new genus, *Carya*, whose name is derived from the ancient Greek name for the walnut. The pecan became *Carya olivaeformis*, a description still in use in 1912 when the Bureau of American Ethnology described it in their *Handbook of American Indians*. But uncertainty continued, and several more changes occurred, including the name *Hicoria pecan* (Marsh.) Britton, which is mentioned in an article on nut breeding by H. L. Crane et al. appearing in the *Yearbook of the United States Department of Agriculture*, 1937. Eventually *Carya illinoensis* was agreed upon and confirmed at the Eleventh Annual International Botanical Congress in Seattle in 1969, in deference to the traditional belief that the pecan orig-

Before mechanization, pecans were harvested manually, with threshers knocking ripe nuts from the branches and pickers gathering them from the ground, as in this Georgia orchard. *Photo courtesy USDA.*

inated in present-day Illinois. Not all plant specialists are satisfied, preferring to believe that as there are millions of pecan trees found along most of the major waterways in Texas and northern Mexico, as well as numerous area fossil findings, this is a more plausible source region.

Fossilized remains of pecan trees are embedded in the lower Cretaceous formation in Lampasas County, Texas. Despite questionable scientific validation, popular pecan lore recounts that Edmond E. Risien, who cultivated early improved pecan varieties in this area in the late nineteenth century, prized a perfectly preserved fossilized pecan that was blasted from its lodging spot at a depth of thirty feet in the San Saba Valley as a well was being established.

The pecan is the largest of the hickories and like all of them has a scaly bark showing grayish brown to light brown under the scales and beneath that a coarse-grained, hard, reddish-brown wood. Between March and May the flowers appear, fertilization occurs, and a new covering of alternate pinnate leaves begins to grow, each up to twenty inches in length and bearing nine to seventeen serrated leaflets.

The nuts grow in clusters of three to eight, each encased in a thin, four-winged husk. Among the cultivated varieties, the size of the nuts falls within a predicted range. However, when the tree is grown from seed, the nut size differs, not only from one cultivar to the next, but also, with some variation, on a single tree. Generally the native nuts range from about three-fourths of an inch to two inches and in shape from long to almost spherical. The pecan shell is a lightish tan, more or less marked by black streaks, and of varying thickness. The kernels fall between golden and cream in color, and some fill the shells more than others.

The appearance of *Carya illinoensis* varies according to geographical region, the density of surrounding trees and plants, and its closeness to water. For example, the fairly low rainfall in the Southwest produces slower-growing, more compact trees with a greater number of branches than those that grow in the more humid habitat of the Southeast.

Pecan trees generally favor alluvial floodplains where they can compete successfully for light and space amid the other plant growth. Their preferred soil is a fine, sandy loam to a depth of about twenty-four inches and below this a sandy clay subsoil that secures the extensive taproot. A permanent, static water table, ideally ten to twenty-five feet below the surface, is the foundation for such soils. The trees are well nourished with the humus and organic matter produced by generations of seasonal inundation and carrying all the major and minor nutrient elements that ensure the tree's long life. Microscopic air spaces in such soil will absorb and hold water from the overflow of rivers and streams and encourage healthy root systems. In the native range, wild pecans crowd the river and stream banks on some eight thousand miles of waterways winding through Texas alone.

In addition to the flowing water, rain (replaced by irrigation in arid areas beyond the native range) is needed to launch spring shoots and leaves, and the growth of the nuts between April and October. Once the nut growth is clearly under way and the frequency of freezes diminishes, the trees are less susceptible to the ferocity of spring storms. But the timing must be right, for such storms, or even prolonged rainfall, can damage the spring flowering and result in a sparse nut crop. Brief flooding is beneficial, but pecans "will not stand wet feet," and few pecan trees grow south of Natchez, Mississippi, where the land has very heavy soil and is subject to long overflow and standing water. At the season's end, heavy rain in the fall can limit the harvesting time, and natural groves may lose a whole crop to flooding.

The pecan has a genetic strategy built in for cold tolerance, as the leaves are killed by the first fall freeze, which may occur around mid-November. During the winter the tree becomes dormant, resting until after the last freeze occurs toward the end of March, at which time the new leaves feel safe to unfurl. The temperature pattern is all-important for the development of a good crop of nuts. While in the native range pecans are adapted to the weather's vagaries, improved varieties are more finicky. The preference is a summer mean temperature ranging between 75°F and 85°F with little variation between the day and night. In the winter, trees may require at least four hundred hours with the temperature falling below 45°F and a mean range of 45°F to 55°F from December to February.

One botanical distinction in *Carya illinoensis* is the tendency to masting, or alternate bearing, producing a good crop of nuts only every other year. At one time this was believed to coincide with years of slower growth, indicating the tree's need to rest. In fact, the tree grows more during the heavy-crop seasons, and the masting is probably an attempt to starve insect pests. The most vexing pest, the pecan weevil, knows how to stay dormant during the "off" year and awaken hungrily when the tree begins to grow again. Pecan growers, challenged by the masting, continue to seek varieties prepared to shed the distinctive, costly trait.

# ▪ HISTORY

The pecan established a botanical niche in North America millions of years ago, but it was probably less than ten thousand years ago that it was adopted, and later adapted, by the earliest settlers. For centuries it was a seasonal item in the American Indian diet. When the Europeans began to explore the continent, they were introduced to the pecan, and some found it so good they questioned the possibility of growing the tree in their own countries. In the nineteenth century a new step was taken when attempts were made to improve the nut by grafting shoots onto an established tree. As the techniques for this venture progressed, a second history of the pecan was initiated. Whereas until this time seedling pecan trees could be planted outside their native range, the improved varieties made extensive orchard crops a reality in areas previously deemed unacceptable due to problems inherent in the climate or terrain.

In the Devils River area of Val Verde County, down where the Rio Grande separates Texas and Mexico, pecan seeds and leaves have been recovered from Baker's Cave alongside human artifacts. They lie in strata dating from 6100 B.C. to about 3000 B.C. but are not found in the Golondrina hearth, where the first human occupation was about 7000 B.C. Whether humans brought the pecan to this location between 7000 and 6000 B.C., or the date indicates the natural extension of the species' range at that time, or whether there are still missing pieces of the archeological jigsaw, is a question yet to be answered. All one can say is that the pecan may well have been known in the area eight thousand years ago.

In *The Beginnings of Agriculture in America* Lyman Carrier contends that for several weeks each fall the pecan groves of Texas and the lower Mississippi Valley provided the main source of food to the poorer tribes of Indians. The wiser and more thrifty among them gathered the pecans and other nuts in the region. The nuts were shelled and pounded, then dried to be used as a seasoning flour for their gruels and maize bread or, sometimes, as liquid *powcohicoria*, fermented for a ceremonial intoxicant. Carrier believed that, despite popular mythology, itinerancy was no more representative of Indian lifestyle than the "tramps and scalawags in our own society" and that most Indians relied largely on agriculture to meet their dietary needs. But between the extremes of permanent settlement and itinerancy lie degrees of transhumance, established patterns of seasonal migration like those that still prevail in mountains and deserts of the Middle East, Asia, and South America.

As the Indians traveled, they likely planted pecan nuts around their campsites to provide "grubstakes" for subsequent stays and for their descendants. Many canoed along the Mississippi and its numerous tributaries, extending the pecan range over a vast network of waterways. Indians of the southern plains transported the nuts by land. Each winter, the Mescalero Apaches took their bison herds from the mountains of New Mexico to Colorado and as far as the Concho River in Texas. There they subsisted by hunting and gathering, a practice continued into the nineteenth century when camps were pitched along the San Saba, Pedernales, and Llano rivers. Pecans were not only eaten but also traded, and probably taken back up into the mountains. Later, the Jumano Apaches took nuts to the explorers along the lower Rio Grande Valley, and supposedly Iguace coastal Indians traded pecans with the Spaniards as the explorers forayed north up the Colorado River.

Explorers marveled at the abundance of pecans, identifying them simply as nuts or walnuts. Cabeza de Vaca with his party, and sometimes as a captive of Indians, crossed the coastal strip between Galveston Island and the Guadalupe River and on farther west and south. A journal entry for 1533 recalls stopping at "the place of which we had been told, to eat walnuts. These are ground with a kind of small grain, and this is the subsistence of the people two months in the year without any other thing; but even the nuts they do not have every season, as the tree produces in alternated years. The fruit is the size of that in Galicia, the trees are very large and numerous."

Lope de Oviedo, a member of Cabeza de Vaca's party, wrote: "There were on the banks of the river many nuts, which the Indians ate in their season, coming from twenty or thirty leagues round about. These nuts were much smaller than those of Spain."

Less than a decade later Hernando de Soto, trudging across the humid reaches of the Mississippi, found the pecan. He and his party had left the land of the Chickasaws to make their way up the great river. Reaching the area known as Little Prairie, they envied the delectable diet there for the picking, including pecans, mulberries, and two varieties of wild plums.

In Texas toward the end of the seventeenth century, Antonio de Mendoza and Alonso De León found themselves in pecan territory. Mendoza crossed the region of the middle Concho River to its junction with the Nueces in 1683–1684 and continued eastward to the Colorado River near its junction with the main Concho River. On several occasions he recorded seeing groves of pecan and live oak trees along the watercourses. When De León marched from Texas to Mexico in 1689, his path took him by the Nueces River, Atascosa Creek, and the Medina River.

While the Spanish staked their territorial claims on the southwestern fringes of the pecan belt, French explorers, trappers, militia, and eventually settlers dug in along the Mississippi and its tributaries. Not long after André Pénicaut and Xavier Charlevoix described the laden pecan trees, Antoine Simon Le Page du Pratz supplied a gourmet twist in his history of early Louisiana. He recorded an early cultural blending when he noted "a kind of very small walnut that would be taken at first glance for a filbert, since they have form, the outline, and the likewise thin shell, but internally they are shaped like the walnut; they are more delicate than our own, less oily, and with a flavor so fine that the French make 'pralines' of them as good as those made of almonds."

This was a compliment indeed. The praline, named for the French marshal César, Comte du Plessis-Praslin, who believed almonds would be more digestible if coated in sugar, was part of the French cultural heritage. It was but a small step to adapt the confection by substituting native-grown pecans for almonds. When Étienne de Boré succeeded in developing a way to granulate sugar a few years later, pralines became not only a treat but a regional fixture.

Until the Louisiana Purchase began to stabilize the region in 1803, the Mississippi tributaries were pulled like taffy by the Spanish and French explorers. Antonio de Ulloa, a Spanish traveler and writer, visited the area drained by the lower tributaries of the Mississippi in the early 1770s and commented: "Two other kinds of trees are found there which appear to be peculiar to that country. One of these they call Pecanos, which is a kind of walnut of greater body than those [walnuts], but in wood and leaf very similar. The fruit is in taste similar to that of the walnut, more delicate and finer, with less proportion of oil. In form it is different, and resembles dates, being in size almost the same or a little less. The shell is thin and smooth and without the roughness which the walnut has."

Although enjoyed by both the French and the Spanish,

pecans were unknown to the British colonists who set up home on the coastal strip bounded by the Appalachians. But other Europeans headed west beyond the mountains from the middle of the eighteenth century on their various missions. As they headed into terra that was still virtually incognita, several of them thought to slip in a few words about pecans. The nuts were granted immortality by Reuben G. Thwaites, who dedicated as much energy as the explorers themselves, if in slightly more comfortable surroundings, to assemble thirty volumes of their reports in his *Early Western Travels: 1748–1846*. He reproduced their accounts as closely as possible, just turning over letters that had slipped sideways and adjusting other "typographic and orthographic peculiarities."

In many cases interest in the accounts was equaled by the authors' reasons for traveling. André Michaux, grief-stricken when his wife died in childbirth, turned to the pursuit of botany and left his home in Versailles, France, with a "desire to study plants in foreign countries." He visited England and Persia before he "herborised on the bank of the Mississippi." During Christmas week in 1795 he camped with his son opposite the "mouth of the River Cheroquis or Tenasse" and passed an "extensive swamp on the North West side boarded by Pekan—Nuttrees" as he entered the Cumberland River mouth six leagues from Fort Massac in the Illinois Territory.

John Bradbury, a Scotsman commissioned by the Botanical Society of Liverpool, traveled from his St. Louis base between 1809 and 1811 carrying out research into plant life in the United States. We can hear the British sniff as he noted that along with the other nuts pecans "furnished abundance of food for hogs."

John Woods crossed the Atlantic to Baltimore and brought his family and goods over the new National Road to Wheeling before flat-boating down the Ohio River to Shawneetown and walking to the settlement of English Prairie in "Illinois Country," as he termed it. The pecans growing profusely in the creek bottoms contributed to his wondrous new happiness.

A more informed naturalist than most commercial travelers, Josiah Gregg, a Santa Fe trader, made eight trips across the great western prairies in the 1830s. Between the Arkansas frontier and Cross Timbers he recognized amid the oak trees many streams bordered by elm, hackberry, pecan, ash, walnut, mulberry, cherry, persimmon, cottonwood, sycamore, and birch.

Edmund Flagg, a prominent early American novelist and dramatist, left his Maine birthplace to become, in turn, a journalist, a lawyer, a consul in Venice, and superintendent of statistics in the State Department in Washington. As a young man he sent articles on his travels to the *Louisville Journal.* One time, leaving St. Louis with evident pleasure, he looked back ecstatically from the banks of the Mississippi at "the closest thing to a cosmorama . . . the American Bottom . . . a tract of country which, for fertility and depth of soil, is perhaps unsurpassed in the world." Valuing effect more than accuracy, he described crossing the nearby Cahokia Creek, "threading through a grove of the beautiful pecan, with its long trailing boughs and delicate leaves."

While Thwaites covered much ground from his desk, many others were eager to see for themselves. John Bartram, an English botanist living in Philadelphia, was champing at the bit to visit the Ohio River, after glimpsing a sampling of unfamiliar flora. "In about two weeks I hope to set out to search myself," he wrote a London friend and fellow botanist, Peter Collinson, on August 14, 1761, "if the barbarous Indians don't hinder me (and if I die a martyr to botany, God's will be done; His will be done in all things)."

He bravely approached territory caught up in the final stages of the French and Indian War and traveled to Pittsburgh via Fort Duquesne, recently captured by General Forbes and Colonel

Bouquet and renamed Fort Pitt. By December he was back and shipping off a box of his finds to Collinson. The response came quickly: "I really believe my honest John is a great wag, and has sent me seven hard, stony seeds, something shaped like an acorn, to puzzle us; for there is no name to them. I have a vast collection of seeds, but none like them. I do laugh at Gordon, for he guesses them to be a species of Hickory. . . ."

Perhaps slightly miffed, Bartram wrote back: "The hard nuts I sent were given me at Pittsburgh by Colonel Bouquet. He called them Hickory nuts. He had them from the country of the Illinois. Their kernel is very sweet. I am afraid they won't sprout, as being a year old."

In an article that appeared in the *American Naturalist* in 1879, Frederick Brendel suggested the pecan was introduced to easterners a little later, after the Treaty of Paris was concluded, when "by chance some fur traders brought a small sample of the nuts to New York."

By 1772 William Prince had planted thirty pecan nuts in his nursery at Flushing, New York. Ten grew, and he made a tidy profit selling eight of the seedlings to customers in England for ten guineas apiece.

Thomas Walter owned a plantation in St. John's Parish on the Santee River in South Carolina where he cherished an extensive collection of plants. With true patriotism he sent many "home," and a number were preserved in the British Museum. In his *Flora Caroliniana*, published in 1787, he described the foliage of the pecan tree, adding ruefully, "fructus non vidi"—"the fruit I have not seen."

Not only foreigners were smitten with the pecan. The French botanist G. L. M. Dumont de Courset recorded hearing from his brother, who served in George Washington's army, that the general was forever munching on pecans and always had some in his pocket. Later, at Mount Vernon, Washington planted "several poccon or Illinois nuts" that had been sent to him.

Thomas Jefferson made several entries on the pecan in his garden book, and repeatedly wrote asking friends and relatives to satisfy his craving. A homesick note from Paris begged, "Procure me two or three hundred Paccan nuts from the Western country . . . they should come as fresh as possible, and come best I believe in a box of sand."

He sent a letter to William Claiborne from Washington in 1786: "I thank you for a bag of peccans lately received from you. If you could think of me in the autumn, when they are fresh, they will always be very acceptable, partly to plant, partly for table use. . . ."

In 1780 he planted pecans at Monticello, but they were still not bearing fruit by 1801. He wrote anxiously to Daniel Clark, Jr., in New Orleans, who replied, "This must be owing to their being planted in too elevated or too dry a soil as they bear in this country in ten or twelve years, and the trees in their natural state are I believe always found in the River Bottoms and in places occasionally overflowed in the annual rise of the river."

The climate was not always to blame. Thomas Mann Randolph confessed a different kind of mishap at Monticello: "The Pacans have not appeared as yet. Thinking that they would not bear transplantation I took the liberty to place them partly at each side of the new way leading from the gate to the house & partly in the garden. Several of those in the garden were destroyed unluckily by the hogs before it was enclosed."

Not long after Jefferson and his relatives struggled to grow the pecan at Monticello, first attempts to grow pecans with scientific precision were taking place in Edgefield, South Carolina, where Abner Landrum was experimenting with budding several plant species in 1822. "The pecan (*Carya olivaeformis*) did not appear to take so well as the walnut but my trials were made rather later in the season." He tried again: "I have, this summer, budded some dozens of the pecan on the common hickory nut, without a single failure as yet; and some of them are growing finely."

At the end of the 1830s the Cherokees were forced westward and resettled close to the Creek tribes on the Arkansas River at the mouth of the Verdigris. As they cleared the luxuriant growth of trees to provide crop land, they left groves of pecans, which they valued as a trading item.

In 1846 the Creek agent, Colonel James Logan, submitted in his report: "The sale of Pecan nuts, the trees bearing which abound in the rich bottom of the water courses, is of considerable importance to this class. It is estimated that the quantity sold to the different traders during the last fall and winter amounted to between 9 and 10,000 bushels, the price for which was from 50¢ to $1. per bushel, and was generally bartered for necessary articles of Clothing, Sugar, Coffee, & Salt, &c, besides a large quantity was no doubt used for food."

Whether or not he had learned of Abner Landrum's efforts, this same year Dr. A. E. Colomb of St. James Parish, Louisiana, tried to propagate his pecans by grafting. His attempts failed, but not his enthusiasm. He cut shoots from a favored tree on a neighboring plantation and took them to his friend Jacques Telephere Roman, who had recently built the Oak Alley plantation on the west bank of the Mississippi some sixty miles above New Orleans. The brother of Louisiana governor André Roman, Jacques Roman preferred hunting, fishing, and developing his sugarcane fields to political socializing, but at the sugarhouse parties he held each year when the grinding season ended, his family and guests enjoyed dipping strings of pecans from his trees in *cuite*, the thick sugarcane syrup.

Jacques Roman passed Dr. Colomb's shoots to Antoine, his slave gardener. Antoine's dexterity and patience led to the first successful pecan grafts, but his full identity is lost. As a slave, he had no last name. The county census roles do not even dignify him as Antoine, granting him only a number, one of many whose faded hand lettering can barely be deciphered. Others soon copied his efforts, selecting traits to improve the taste, hardiness, and adaptability of the nuts. Pecans, no longer growing wholly at the whim of nature, were soon on the way to becoming a profitable modern enterprise.

Early varieties of pecans were graded using this apparatus. *Photo courtesy USDA.*

# ▪ CULTIVATION AND IMPROVEMENT

During the nineteenth century the United States evolved from a new state with scattered settlement along the eastern front to a thriving population of immigrants who crisscrossed the nation's wagonways and later the railroads in their eagerness to find a livelihood. Once the industrial revolution crossed the Atlantic, the cities of the Northeast mushroomed, providing hundreds of thousands of manufacturing jobs. Other settlers chose farming, a more familiar lifestyle, settling the countryside where, as the century progressed, advances in technology and agricultural innovation transformed the American landscape. The story of pecans edged into the realm of science, moving beyond a casual trading item into a regulated orchard crop industry. Varieties or cultivars were selectively grafted from the seedling pecan trees crowding the river bottoms throughout the native growing area, allowing standards to be set for cultivating, producing, processing, and marketing. Those involved in the changes matched the mosaic of American settlement. The Choctaws, living between the Alabama and Mississippi rivers, and who many believe excelled every other North American tribe in their agriculture, probably led other tribes in planting carefully chosen pecans long before European settlement. Following their efforts, French, English, German, and, not least, African Americans became involved.

Although people saw that like most orchard crops pecans did not grow true to seed, it was some time before the growers had a clear grasp of grafting techniques and could be fairly sure of what to expect as the nuts ripened each fall. They built on the success or failure of each trial to produce pecans suitable for the environment in which they grew. From the earliest grafting attempts in Louisiana, pecan culture began to shift beyond the southeastern states, moving gradually westward. New cultivars were developed, then abandoned, especially in the torrid climate of the Southeast, where insect pests and plant diseases frustrated efforts. Over the span of two or three generations, the work was concentrated in half a dozen areas, and among the many who experimented and produced successful varieties, a few individuals made particular discoveries, sharing new grafting techniques, finding especially promising trees, and later cross-breeding the cultivars to introduce still more improvements.

Despite the improvements, extensive seedling crops are harvested profitably even today. They are gathered from trees that have grown without human interference, from nuts that have fallen to the ground, germinated, and then pushed up through the soil on the same land that has nourished them for thousands of years. The nuts tend to be smaller than the cultivated ones—not necessarily a bad feature—but their shells are generally thicker, and they may be hard to reach and shake from the trees. These are the nuts cracked under foot on a Sunday ramble, the happy taste of countryside before the picnic proper begins. There are also planted seedlings that grow in managed orchards and retain their native hardiness. Seedling trees live in harmony with the flora and fauna that share their space. They are stable and have achieved a balance within their ecological sphere. However, this advantage also limits them to their range. If they are planted far from familiar territory, they fall victim to other environmental conditions and are unable to bear productively. Seedlings are slow to reach maturity and may not bear for as long as fifteen to twenty years. Several improved varieties, cultivated in areas beyond the native belt, may produce nuts within six or seven years, a considerable benefit to growers who have invested heavily and need to harvest as soon as possible. It is rare for a cultivar to reach ten years without bearing.

The pecan has two sets of chromosomes, producing pistil and stamen flowers on the same tree. The trees blossom in the early spring, with the male flowers appearing at budbreak, and the female flowers as the new growth reaches about six inches. Some varieties show protandrous behavior, in which the male flowers come into bloom first, shedding clouds of pollen as they split open from anther sacs held on their yellowish-green catkins. Other pecans are protogynous. Here, the female flowers on the tree are receptive to fertilization before the pollen dehisces. Providing the weather is not too hot, or dry or windy, receptivity will last for several days. When pecans are grown for commercial production, the growers must be aware of which varieties play which role. For example, the Cheyenne and Pawnee, two western varieties cloned in Texas, are protandrous and will pollinate the Wichita, Kiowa, and Mohawk cultivars. After ten days or so reverse pollination of these same trees occurs. But the habit of dichogamy, the characteristic which determines whether the male stamens or female pistils on a tree will mature first, varies according to the actual location of the trees, so that, for example, the Western Schley, Desirable, and Barton are protandrous or protogynous in different places. Recognizing the dichogamous trait in specific places is a key factor in successful pecan production. As the wind serves as the pollen's carrying agent, the trees have no scent or nectar and no bright colors to attract insects. Growers plant at least two varieties, with a minimum of 20 percent of the trees in one variety to ensure an adequate supply of the fertilizing pollen. Although self-pollination is minimal and the pollen may waft over a large area, it is judicious to plant protandrous trees not farther than eight orchard rows from the protogynous trees. Instead of allowing natural or sexual reproduction, all cultivated pecans are propagated asexually, by one or another grafting technique, so that each clone is genetically stable. Seedling trees are also big commercial producers, but they are characteristically adapted to the terrain where they grow and serve to retain a wider gene pool for such time as new traits are sought.

The pecan has one especially challenging shortcoming: it is prone to masting, or alternate bearing. As researchers work to improve pecan quality, expand the geographic range, and develop better resistance to environmental hazards, they must try and reduce the tendency of the trees to follow a prolific harvest with a meager one. The Mobile, Tesche, and Frotscher varieties are handicapped by this problem, and several others are prone to overbearing, with excessive weight that can damage the tree. In addition to regular and adequate production, the *Texas Pecan Handbook* (published by the Texas Agricultural Extension Service at Texas A&M University) lists as the desired characteristics of improved varieties the kernel quality, tree and limb strength, kernel percentage in each nut, appropriate size (too large a nut can lead to poor kernels; too small may indicate a low yield), a ripening date before the first freeze, pollination type, cold tolerance, minimum chilling requirement, and precocity of tree production (beginning to bear at an early stage).

As pecan cultivation became established, growers increasingly recognized the need to respect, rather than defy, the environmental advantages and discrepancies in the region where they worked. It became evident that cultivars could be separated into two main groups, those which grew successfully in the East and the others that met the stringent demands of the more arid West. By a nice touch of cultural intervention, the east/west divide follows along U.S. Interstate Highway 35, which stretches from Canada to the Mexican border, passing through central Texas, the world's pecan hub. The varieties grown west of this highway would not thrive on the other side of the divide because of the problems that exist where humidity intensifies the heat. In the Southeast these western cultivars soon fall victim to scab, blight, and "foreign bodies" waiting to devour them. By the same token, the eastern varieties could not survive the aridity and higher altitudes confronting western ones. For example, for close to a century the cultivar Desirable, which originated in Mississippi, has been a great success from that state westward through East Texas and the Texas Gulf Coast, and although its popularity continues to creep west, its range ends well before it reaches the dry atmosphere of West Texas.

In addition to the east/west divide, the cultural range is divided into five belts that extend beyond the native range. This considerably expanded range reveals the benefit of understanding the interaction of soils, climate, plant genetics, and botanical and biological factors that were unknown, or at least unexplained, until the nineteenth century was drawing to a close. Pecan trees can be cultivated for ornamental use north of the native boundary, but generally the nut production has expanded laterally, and there are now impressive plantings from the Atlantic coast to Arizona, with experimental pockets close to the Pacific. The boundaries of the original three belts are defined by the number of growing days each may expect.

An early seminal book, *Pecan Growing*, was written by H. P. Stuckey, who for many years was the director of the Georgia Agricultural Experiment Station, and E. J. Kyle, professor of horticulture and dean of the School of Agriculture at Texas A&M University. They delineated the three accepted belts and added a fourth. Then and now, the southern belt is the most important for commercial nut production, despite severe problems among some older cultivars that are succumbing to insect infestation and scab. Growers in this belt expect the 270–290 growing days many pecans need in order to reach maturity. The northern boundary of the belt reaches from Wilmington, North Carolina, westward to a spot some fifty miles north of Augusta, Georgia, then through Atlanta and Birmingham, Alabama, and continuing southwest almost to Jackson, Mississippi. The boundary moves north across the Mississippi River close to the twenty-fourth parallel and on through Pine Bluff, Arkansas, and McAlester, Oklahoma. On the eastern rim this area is about one hundred miles deep, but it gradually increases to a depth of three to four hundred miles from the coastal boundary just west of Galveston, Texas, as it reaches Arkansas, Oklahoma, and central Texas. When Stuckey and Kyle's book was published in 1925, productivity among the states ranked Georgia in the lead among orchards planted to improved varieties, followed by Florida, Alabama, Mississippi, Louisiana, South Carolina, Texas, and Oklahoma. Growers noted that nut size in given cultivars decreased as the plantings went north through the Piedmont country and neared the Appalachian highlands. For example, in southern Georgia a pound of Stuart pecans runs about forty-five nuts, but farther north the number per pound increases to over sixty. Growers observed, too, that the state ranking was due in part to the fact that some states, such as Texas, had such a wealth of native pecans that they were reluctant early on to invest in grand-scale improvements.

The middle belt has 180–200 growing days a year. In the East it begins at Newport, Rhode Island, travels southward almost to Asheville, North Carolina, then around the Cumberland Mountains almost due north to Louisville, Kentucky. From here it continues on through Vincennes, Indiana, and Bellevue, Illinois, before taking a northern turn to cross the Mississippi at Hannibal, Missouri, makes a southward line around the Ozarks in southern Missouri, then moves again northward through Moberly and St. Joseph and westward in a line with Santa Fe, New Mexico. This belt includes northern portions of Georgia,

Alabama, and Mississippi, the upper Piedmont of South Carolina, the North Carolina Piedmont, coastal Virginia, Tennessee, western Kentucky, southern areas of Illinois and Indiana, and southern and middle Missouri. As the environment allowed seedlings to grow abundantly along the Ohio River, this area lagged behind the South in experimenting to develop improved varieties.

The northern pecan belt lies east of the Rocky Mountains and has 170–190 annual growing days. From Portsmouth, New Hampshire, the boundary runs south close to Cumberland, Maryland, then northward through Pittsburgh and Bradford in Pennsylvania and Auburn, Syracuse, and Watertown in New York. It skirts the western edge of Ontario, Canada, reaches through Detroit to the northern edge of Indiana, and back over to the forty-third parallel, through Grand Rapids, Michigan, and Milwaukee, Wisconsin, before bearing southwesterly to Trinidad, Colorado.

Stuckey added a fourth belt that includes eighty-one counties of West Texas, where the semiarid climate nurtures native groves in the valleys between the high hills and mountains of the region. The edge of this belt rises in Young and Jackson counties and moves southward to San Antonio. The soil is rich and deep from mountain erosion and receives eighteen to forty inches of rain annually. Many Texas varieties grow well in this region, at altitudes between 1,000 and 1,800 feet.

A fifth belt, termed the extreme western or irrigated belt, has now been defined. It includes West Texas, lower New Mexico, Arizona, and California. Commercial expansion here has come about since the 1950s, with a marked increase in Arizona. Most of the land set to pecans is in the area south of Tucson, where the nuts have replaced cotton. The cotton industry withered after new government regulations were implemented in 1949 and the development of synthetic fibers forced rethinking of land use. Several fruit crops were considered, including plums, pears, pecans, persimmons, peaches, nectarines, and wine and table grapes. Pecans were chosen, as they produced more calories per acre than any other possibility except dates, which needed twice as much water. Today technology has made vast orchards not only possible but also practical, among them a five-thousand-acre spread of pecans in the high desert about fifteen miles south of Tucson in the Santa Cruz Valley.

The Arizona operation is a distant cry from the first attempts to make the pecan a predictable crop, and beyond the imagination of early nineteenth-century horticulturalists. The traveling botanist André Michaux, who saw and admired riverbank pecan groves in Kaskaskia, Illinois, in 1819 suggested attempting to graft pecan stock onto the black walnut for commercial production in the eastern United States and in Europe. In 1822 Abner Landrum tried setting pecan shoots on the "common hickory nut." It was almost a generation later that Antoine, the slave gardener in Louisiana, successfully grafted pecan shoots on mature pecan trees, marking the first recorded improvements. By the end of the Civil War, he had propagated 126 trees at the Oak Alley plantation. The first named variety, the Centennial, came from one of these.

The purpose of grafting is to reproduce selected and thus predictable characteristics. The procedure is carried out by taking buds or shoots (known also as scions) from one tree and transferring them to an established tree, or stock, where they will maintain their own genetic makeup while being nourished by the host tree. Scions, or shoots, are cut from a tree having the desired characteristics, then by one or other selected method they are attached to established stock where they can mature and bear nuts. The shoots can be grafted onto saplings or "topworked" on mature trees. Or, instead of scions, growers may choose to graft single buds together with their surrounding tissue. In the early days this was chancy and demanded plenty of patience as well as dexterity.

Edmond E. Risien, a transplanted Englishman working in central Texas, came up with a technique he called "ring-budding." The method made the operation far easier and more reliable, and many other growers gratefully adopted it. In *Pecan Culture for Western Texas*, which he published in 1904, Risien described his technique, for which he used a double-bladed knife and fabric strips dipped in a protective coating of wax. "The buds to be set must be fresh," he wrote; "it were better that they be cut the same day as they are used, and protected from drying out by wrapping them in something damp. . . . While holding together both scion and stock, now mark with a sharp knife across both, above and below the bud on the scion and between the buds on the stock, so that the scion shall exactly fit the space on the stock where the bark was removed. . . ." The knife cut clear around the narrow shoot and loosened half of it before "with a little twist of the thumb and finger, at the same time pressing gently on the bark, the whole ring [was] made to come off." The bud and its surrounding bark had to be inserted within minutes, then the wound wrapped tightly with waxed cloth and left for two weeks. By this time the favored buds would have started to grow, and the wrapping had to be loosened if it had not already burst.

Risien's technique soon triggered improvements. J. A. Evans, a pecan specialist at Texas A&M, introduced the "patch bud" or "modified ring-bud." This was more flexible, as a single patch of bark, measuring no more than four inches square, could be removed with less risk than the ring. The technique enjoyed most success if carried out in the spring when the growing is most vigorous, but buds could be inserted as soon as the bark slipped off easily and until drought or the dormant winter period set in and the bark refused to slip. The bud was attached to a space removed from the rough bark covering the trunks or branches of a tree, and if the bud failed to "take," the tree would suffer no damage. Patch budding is still popular, especially for nursery propagating and topworking on small trees, but the technique has been somewhat refined and is more commonly carried out now in the summer using buds from new growth.

Then field researchers began to experiment with topworking—cutting the top branches off larger pecan trees where the early growth years had already taken place. The transplanted shoots were then integrated into the growth pattern and produced far earlier than when propagated onto younger, smaller trees. When the practice was introduced, trees were cut back during the dormant period in winter to force out especially vigorous sprouts where the shoots could be attached. By this method, pecans were growing within two or three years.

Several orchards were planted in the years following the Civil War, initially with seedling trees. Within a generation the advantages of the improved varieties were becoming recognized, partly due to nursery catalogues singing their praise. Compared with today's orchard size and productivity, the beginnings were humble. There were no more than fifty trees in the first orchards to advertise their wares in Federalsburg, Maryland, and Danville, New York. In Illinois several fifteen- to twenty-acre orchards were planted, and the six hundred pecan trees in Martinez, California, made an impressive spread. Aside from these and similar clusters, the main centers for pecan orchards were in southern sections of Louisiana, Georgia, and Mississippi, northern Florida, and the upper Colorado River area of Texas.

T. H. McHatton, who chaired the Department of Horticulture at the University of Georgia from 1908 to 1951, termed Louisiana the "cradle of improved pecan production." This was where Antoine had carried out the pioneer grafting. It was also home to the one-time largest pecan tree, which stood 107 feet tall, had a spread of 135 feet, boasted an unusually large circumference, and one season produced an estimated three thousand pounds of pecans. It stood in a plantation in Ascension Parish on the west bank of the Mississippi and was dedicated by the National Pecan Association in 1927 to the "yet unborn and future pecan growers of America." But even more, Louisiana won its reputation because of New Orleans. Besides being a good market for orchard crops and planting stock, the city also served as a redistribution point for seedling and improved nuts and planting stock. Many of the southern improved varieties came from nuts purchased in New Orleans, brought there mostly from Mississippi and Texas.

Native trees were growing in tall, natural hedges back around 1877 when Emil Bourgeois of St. James Parish revived the technique of grafting, which had seemingly languished for thirty years after Antoine's painstaking work. Bourgeois took twenty-two shoots from a tree planted in the parish in 1836 and set these as top grafts on seedling trees. As they matured, he propagated young trees for planting in orchards and for sale to nearby planters. Some of the nuts and scions from these trees were acquired by Colonel W. R. Stuart of Ocean Springs, Mississippi, who nurtured them and successfully developed a cultivar that he named Van Deman, in honor of Professor H. E. Van Deman, a respected fruit scientist with the U.S. Department of Agriculture (USDA).

In 1879 Sam H. James planted the first large seedling orchard in Mound, Louisiana, choosing nuts from New Orleans. When the nuts ripened, he was disappointed that they were not true to their origins, so he began to propagate. His patience was rewarded with the Moneymaker, a variety which is still grown in six states and accounted for 2 percent of the total U.S. pecan production in 1990.

William Nelson was the first to propagate pecans commercially in Louisiana. He worked with nurseryman Richard Frotscher, who, in 1885, published the first catalogue to offer improved varieties. The catalogue listed three choices of improved pecans: the Centennial, Frotscher, and Rome.

The Centennial derived from one of Antoine's grafts. In 1876 Hugh Bonzano, the new owner of the Oak Alley plantation, entered nuts from his grafted trees at the Philadelphia Centennial Exposition. They were awarded a diploma and recommended for their "remarkably large size, tenderness of shell, and very special excellence." The name Centennial commemorated this honor. The original Centennial tree was swept by floods from the east bank of the Mississippi in 1890.

The second catalogued variety, the Frotscher, came from a tree planted in a garden beside Bayou Teche in Iberia Parish around 1860. It was first marketed as Frotscher's Eggshell and is still grown quite extensively in Georgia and South Carolina.

The Rome cultivar came from a tree planted by Sebastian Rome in his garden at Convent in St. James Parish around the year 1840. The nuts were enormous, averaging twenty-five to the pound (compared to the more customary forty pecans per pound). Emil Bourgeois took scions from the mother tree and top-grafted them onto seedling trees at Rapidan, his plantation in St. James Parish. He marketed the Rome as Pride of the Coast. For this largest, most eye-catching variety, growers came up with other names to draw attention to it: Century, Columbia, Mammoth, Southern Giant, and Twentieth Century. It was not uncommon for an improved pecan to have more than one name, but the practice could be exasperating. The names were frequently chosen to boost particular attributes or were linked with whomever was responsible for developing them, or someone the developer chose to honor. The Van Deman was a case in point. It was known variously as the Duminie Mire, Mire, Mere, and

Meyer (all to immortalize Duminie Mire, who had secured the nuts from a neighboring plantation to plant the first tree), the Bourgeois (for Emil Bourgeois, who propagated scions from it), and also, more poetically, the Paragon and the Southern Beauty. In 1904 the National Nut Growers' Association, perhaps wearying of the confusion, adopted the nomenclature practice of the American Pomological Association, to allow each cultivar a single name and to ensure that each name refers to only one variety. Since 1934 it has been possible to patent a cultivar under the U.S. Plant Patent Act. This gives the owner jurisdiction over the propagation and sale of trees and scions, although the patent does not automatically reflect the merits of the cultivar.

In Louisiana, the pecan was popular as an ornamental backyard shade tree and for home orchards. After severe frosts in the winters of 1894–1896 destroyed citrus orchards, the resilient pecan became commercially appealing, and more improved and seedling trees were planted as backup or secondary orchard crops. By 1902, fifteen improved Louisiana varieties were available to growers; many of these came from St. James, Madison, and St. Charles parishes.

The greatest variety of cultivars originated in Mississippi, especially in Jackson County, where early pecan cultivation was centered around Pascagoula, Scranton, and Ocean Springs. In her unpublished dissertation on the region's pecan culture and history, Jean Richardson Flack suggested that Ocean Springs had several advantages. It was surrounded by native trees, close to New Orleans and available Louisiana stock, and targeted by German immigrants who were sophisticated in European agricultural practice and saw a future in pecans. Several of the growers there left a legacy of pecan varieties which are still popular more than a century later.

About 1875, James Moore, a blacksmith in Ocean Springs, sold pecans to Colonel W. R. Stuart, who had retired there at the age of fifty-six, leaving a successful cotton and sugar business in New Orleans in order to become a gentleman farmer and pursue his interest in horticulture. Colonel Stuart's productive retirement years made him a giant in pecan history. He bought a hundred seedlings from Mobile and New Orleans and later propagated several notable cultivars. Five of his best pecans were bought by Peter Madsen, whose property was subsequently sold to a Mrs. H. F. Russell. Remarkably, four of these five nuts became successful cultivars, and one was named the Russell by Charles E. Pabst, the first man to sell grafted stock in Mississippi. A cultivar named for Pabst still holds 0.3 percent of the American market.

The cultivar that immortalized Colonel Stuart's place in pecan history was first offered for sale in 1892. Even at the turn of the century it was recognized as the most widely successful pecan variety yet introduced and tested. More than a century later the Stuart pecan is still the most popular variety in the world, occupying 27 percent of all improved orchard space and accounting for at least 50 percent of the cultivated crop in Alabama, Louisiana, Mississippi, North Carolina, and South Carolina. The original Stuart pecan was brought to Mississippi from Mobile, Alabama, and planted in Pascagoula by John R. Lassabe. Within fifteen years the tree's crop averaged 140 pounds, and a bumper season in 1892 produced a 350-pound yield. Charles Cruzat, who leased the land on which the tree grew, pocketed a tidy profit when he sold the Stuarts at a dollar a pound. (Cruzat's little son helped the family further when one of just six nuts he bought for fifty cents in New Orleans was chosen for the crossbreed Jewett, named in honor of a Colonel Stephen Jewett of Crosby, North Carolina.)

The Stuart's rapid success was attributed to its large size, which appealed to the in-shell trade, its fair resistance to scab, the moderately high nut yields—and very good promotion. Today it yields on average twelve to fifteen hundred pounds per acre each year.

Theodore Bechtel began working for the W. R. Stuart Pecan Company in 1899 and later bought land and started his own nursery at Ocean Springs. One of the seedlings he propagated became the Success in 1903 and today is still a considerable achiever. Ten years later he introduced the Candy, an unusually small cultivar, suitable for purchasers requiring a decorative pecan for the baking and confectionery market. Where seedling orchards continued to maintain their popularity and high productivity, growers raised their eyebrows at the extraordinary idea of deliberately cultivating smaller varieties when they could be harvested so plentifully and easily from unimproved seedlings.

Charles Forkert, one of the German-born Mississippi growers, was credited with developing the first controlled crossbreeds. From 1903 until his death in 1928, he planted hand-pollinated seeds that led to the Admirable (Russell x Success); Dependable, Desirable, and Superdesirable (all Success x Jewett); the Forkert (Success x Schley); and Favorite (Russell x Success). He did not live to see the spread of his greatest triumph, the Desirable, which is today the preeminent variety in the Southeast, but in 1925 he sent scions to the United States Pecan Field Station near Albany, Georgia, and ensured its future. The Desirable takes longer than some varieties to reach bearing, and its susceptibility to scab is increasing, but it is a prolific, constant bearer with an easy-crack shell and has a delicious, plump, straw-colored kernel. It is grown in nine states, with the second greatest improved acreage and third highest productivity, totaling 10.9 percent of national production.

Beyond the eastern boundary of the native growing range, but where maybe a scattering of wild trees grew, Florida became an early entrant in commercial pecan growing. In 1848 when a certain John Hunt returned from military duty, most likely serving in the Mexican War, he established a seedling orchard at Old Bagdad, near Gainesville, with nuts like those he had enjoyed on his travels and likely gathered along Texas riverbanks as he made his way home. His orchard became the nucleus of several other ventures in northern Florida and southern Georgia. It was the source for such entrepreneurs as the physician J. B. Curtis, who planted an orchard at Orange Heights almost forty years later. Dr. Curtis developed many pecans from his seedling trees, benefiting the local market both with the nuts that he produced and with improved pecan stock that was fully adapted to the area.

The icy winter that devastated orchard stock in Mississippi did not spare the citrus crops of Florida. In a small but important center near Monticello, H. S. Kedney, whose orchards were severely damaged by frost in 1886, planted four thousand pecans from Texas. Several other growers were tempted to try the new crop once they recognized its resilience, and within a few years pecans were planted as backup crops in anticipation of citrus damage and eventually became an integral part of the local agricultural picture.

Pecan culture in Georgia took a different path. During the 1880s a few seedling pecan orchards were planted in Thomas County, not far from the large orchards across the state boundary in Florida, and others were established near Savannah. An orchard of five hundred trees was planted close to Albany in 1887 and had increased fivefold by 1889. At this date the entire state had only ninety-seven acres planted in pecans. Louisiana, meanwhile, already boasted a thousand planted acres and Mississippi twice that number. Traders passing through Albany on their way north bought pecans not sold on the local market. Satisfied by their purchases, they sent repeat orders, and with little warning commercial pecan ventures took hold.

Scarcely were the few orchards planted and the growers dedicated to grafting good stock, when entrepreneurs saw that selling small orchards might be even more profitable than the pecan crop. With an eye to a fast dollar, agents triggered a speculative land rush. By 1900 many had entered the "pecan game." Old cotton plantations were revitalized as pecan orchards were hastily planted. In some areas land prices quadrupled when buyers, mostly from the East and Midwest, were hired by exaggerated promises based on unfounded estimates of yields and profits. One promoter claimed that a five-acre pecan grove would net $2,500 a year with minimal upkeep and begin producing papershell pecans in a couple of years, enough to keep the average family in comfort. A five-hundred-acre orchard in Albany was sold through a Chicago agent in five-acre lots, realizing a considerable profit for the first owner. Over and again hundreds of acres planted in pecans were sold to speculators in units as small as a single acre. Between 1909 and 1919 pecan production in Georgia grew from 354,046 pounds to 2,544,377 pounds, and all of it well to the east of the native range.

Foreseeably, the bubble burst as the Depression years loomed. Although some speculators had become rich, many had failed. For several years afterward the orchards went without adequate management and suffered deterioration and damage that seemed irreparable. Then came a movement to restore them, and Georgia rose to first place in productivity. But today the technology that has made it possible to grow pecans successfully in arid states is moving the industry westward. Despite the Stuart's still outpacing all other varieties, there are several cultivars now proving their suitability on the western edge of cultivation that will soon topple it from leadership.

# ▪ TEXAS: A CASE STUDY

Texas holds a special place in the story of pecan culture. The pecan grows wild in 150 counties, and commercial orchards have been planted with improved varieties in 200 of the state's 254 counties. There are pecan trees alongside eight thousand miles of rivers and creeks, especially beside the Colorado, Nueces, San Antonio, Guadalupe, Brazos, Trinity, and Red river systems. The Colorado alone has several major and minor tributaries: the San Saba, Concho, Llano, and Pedernales rivers contribute to its drainage system, and the Peach, Caney, Jones, and Blue creeks feed into the lower reaches. The state has more than fifty pecan orchards larger than seven hundred acres and, at the other end of the scale, three thousand enterprises that are run by single families, with seasonal help during harvesting. Even today over 70 percent of the pecans harvested are from seedlings, standing on more than 700,000 acres of land across the eastern and central parts of the state—and this despite the fact so much acreage is covered by the improved western varieties developed by the Texas-based USDA Pecan Research Station.

There are other distinctions in the history of the state's pecan development. The technique of ring-budding, a benchmark for widespread propagation, was introduced there, and researchers in West Texas found a way to combat rosette, which, until Louisiana researchers identified the cause as treatable zinc deficiency, had been a perennial pecan scourge. The reports from early pecan-association meetings show dedicated, patient growers sharing their views, their wisdom, and the results of their experiments, all with gentle humor.

Once Europeans began to settle Texas in the 1820s, it was not long before their activities reached beyond the confines of their homesteads and farms. There were those who helped win independence from Mexico and achieved military honor, and others who expanded their original holdings and, introducing European methods, diversified their produce. Some became wealthy traders. Still others became folk heroes, their aura untainted by prosperity. General Ben McCullough proved himself on the battlefield, helping to establish the Republic of Texas in 1836, and was elected to the Texian Congress while still in his twenties. In addition to his political and military eminence, he knew from a young age how to take advantage of what Texas had to offer. Together with his brother and two friends, he camped as a youngster on the Guadalupe River bottom and built a flatboat to hold the pecans that lay there. They sailed downriver to test the river's navigability and their own business acumen. A trader bought the nuts from them at Pass Caballo on Saluria Island, which was a stop on the way to New Orleans. Well satisfied, the young men gave the boat to people living on the island.

It was quite commonplace in the mid-nineteenth century to transport pecans downriver. The German botanist Ferdinand von Roemer saw wagonloads of pecans gathered in the fall and brought to Houston for shipping to the northern states. By 1857, 200,000 bushels were exported annually from the state, and in 1871 pecans in some parts of Texas were worth five times as much as cotton. There was endless squabbling about the comparative merit of leaving pecans where they stood or cutting them down and setting the land to cotton. The controversy drew an impassioned outburst at the 1907 Texas Nut Growers' Association when E. W. Kirkpatrick, a crony of the late Governor Hogg, censured the practice of clearing pecan trees for the plantation crop. "We have hurt ourselves and our soil raising cotton, and we have cut down trees that were worth $500 . . . and dragged our women and children down, and ignored and impoverished them, and reduced them to almost slaves to pick cotton and sell it for 5 cents a pound. . . . We have pursued the wrong course in neglecting the pecan . . . a more profitable thing than the walnut groves of California. It excels the olive, orange, the lemon, the peach, the apricot, the prune, and the grape: it excels everything."

In the early days, growers debated how much to invest in developing improved varieties when the state had such an abundance of fine pecans free for the picking. Their debate did not interfere with appetites. When, in 1914, the *Country Gentleman* ran an article on pecans, the author, James H. Collins, boasted: "Tyler is a Texas town with about 12,000 people who eat a carload of pecans every year. If New York ate pecans at the same rate, it would consume our whole crop. Yet we know of many families whose per capita consumption of pecans is five to ten times the Tyler average."

Nor did growers allow themselves to fall short on learning about new developments in the field. Texas A&M University initiated a course in pecan culture in 1910. It has been taught continuously since then, with only three teachers in nearly a century. While many careers were launched from this course, and the progress of the state's pecan industry has been closely documented, there are still mavericks who chose not to brag too accurately on their attainment and became the bane of census takers. One successful but cagey grower, P. K. DeLany, bought his farm alongside the Guadalupe River in central Texas in 1895 for $8.75 an acre, thumbing his nose at those who predicted he would be unable to raise crops on such low-lying land. At first, pecans were bringing just one and a half cents a pound at the local store in nearby Seguin, and as the railroad did not come through, his crop

Pecan nuts grow in clusters of three to eight, each encased in a thin, four-winged husk. *Photo © 1994 by Paul M. Montgomery.*

The pecan tree (*Carya illinoensis*) thrives naturally in alluvial floodplains, where it can compete successfully for light and space amid the other plant growth; it typically grows to a height of 100–180 feet and has an average life span of 70–75 years. *Photo © 1987 by Paul M. Montgomery.*

Nuts ripen in the husk, which breaks open in fall to reveal the shell, with the kernel inside. *Photo © 1994 by Paul M. Montgomery.*

Pecan varieties illustrated in the *Yearbook*, U.S. Department of Agriculture, 1907.

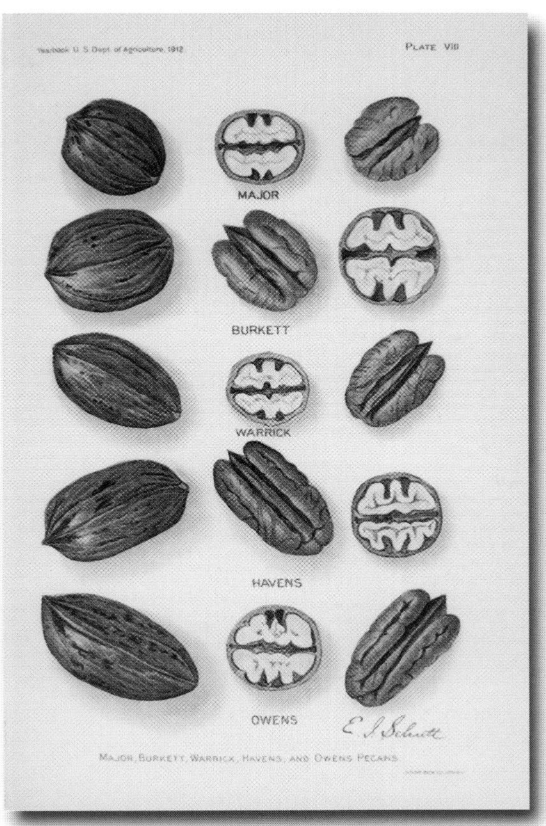


Pecan varieties illustrated in the *Yearbook*, U.S. Department of Agriculture, 1912.

For pecans to thrive, they need a suitable climate, adequate water supply, protection from predators and disease, and well-nourished soil. *Photo © 1986 by Paul M. Montgomery.*

Squirrels are among the pecan's most appreciative consumers—and most aggressive predators. *Photo © 1994 by Paul M. Montgomery.*

Pecans are used in a wide variety of dishes, especially candies, pies, and fruitcake. *Photos © 1994 by Paul M. Montgomery.*

had to be taken into Luling for shipment to points beyond. Then transportation improved, supply increased the demand, and the price rose. When asked how many trees grew on his 417 acres, he would reply, "They've never been counted, but that's what the government keeps trying to find out." In a good year, he harvested around 200,000 pounds of pecans.

As recently as 1987 the U.S. Department of Commerce Census of Agriculture did not supply figures for all Texas counties, perhaps in deference to those larger operations where the grower's identity would become obvious if other growers did not turn in their figures. Bad weather conditions, an epidemic, and the continuing frustration of alternate bearing might all detract from the trend of increased productivity. In the past growers had also a more human problem to deal with. The fledgling industry was rife with intrigue, and it was not uncommon for growers to "hide" their favorite trees, the ones they had earmarked for special distinction. Richard Govett, one of several transplanted Englishmen among early Texas pecan growers, introduced a promising cultivar in 1922 from an excellent tree on his land close to Seguin on the Guadalupe River. It was "protected by bobwire entanglement almost as heavy as on a battlefield. Surrounding trees were also so protected and no one ever had the exact tree pointed out." The exclusive right to all budding and planting material from this tree was granted for ten years to a man who paid $1,000 for the privilege, and a California nurseryman bought the right to propagate the tree on condition the name Govett was not used, to protect local sales competition.

In San Saba County, some 130 miles north of Seguin as the crow flies, Edmond E. Risien developed the ring-budding technique to propagate pecans. As a result, an increased supply of stock became available on the market, and the nursery price on trees was reduced from two and a half dollars apiece. Risien was born in Kent, England, and settled in central Texas, where he

British-born Texan Edmond E. Risien made significant contributions to the development of cultivated pecan species in the early twentieth century. *Photo courtesy USDA.*

earned a living as a cabinetmaker and devoted as much time as he could afford to his orchard. He offered five dollars for the best pecan brought to his cabinet shop. The winning nut came from a tree that stood near the confluence of the San Saba and Colorado rivers. He bought the land, and the tree became his laboratory. It took some time before much could be accomplished, as the previous grower had cut all but one branch from the trunk, so he could stand on the remaining one in order to gather the nuts.

Risien admitted that some propagation attempts were less rewarding than others: "There is not always a congeniality between the stock and the scion, and while it is seldom this way, we have proof that it is so, because we see some unions that refuse to do any good; they don't grow off right and if compelled to grow together, in some ways they are disappointing. Sometimes I think we may just as well let them separate, as is the case with some married couples, but, on the other hand, when we see the scion growing off thrifty all will be well." Despite occasional mismatches among the one thousand nuts he selected and planted from the mother San Saba tree, he laid the foundation for western pecan varieties. His successes included the original San Saba cultivar, the San Saba Improved, Onliwon, Texas Prolific, Squirrel's Delight, and the Western Schley. This last, now known simply as the Western, is one of the mainstays of production in the western states. A protandrous tree, it bears a medium-size pecan weighing fifty-two to the pound—almost 60 percent of the weight is kernel. The Western is second only to the Stuart among all improved varieties.

Risien established a nursery and ran a telephone line from his business to the post office in nearby Rescue so that he could prepare orders received in the afternoon mail for dispatch the following day. Besides the promotional catalogue that showed his enthusiasm and personality as well as his prices, he sent barrels of pecans to children in his hometown in England and to the armed forces during the wars. He also sent samples to Queen Victoria, the poet Alfred, Lord Tennyson, President McKinley, and members of European royalty.

Halford A. Halbert, another early Texas grower, taught school, practiced law for ten years, and brought out a weekly

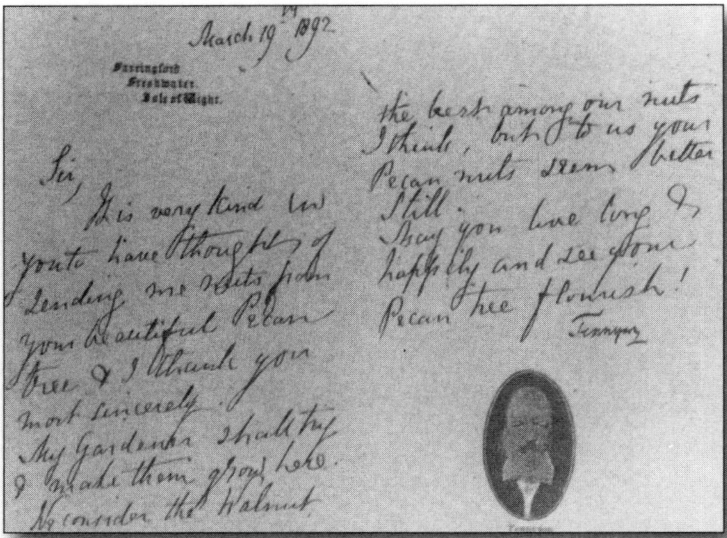
Farringford
Freshwater
Isle of Wight.

March 19 1892

Sir,
It is very kind in you to have thought of sending me nuts from your beautiful Pecan tree & I thank you most sincerely. My Gardener shall try & make them grow here. We consider the Walnut the best among our nuts I think, but to us your Pecan nuts seem better still. May you live long & happily and see your Pecan tree flourish!
Tennyson

Edmond Risien assured fame for his work with pecans by sending them as gifts to notables overseas. From *Famous Paper Shell Pecan Commercially Grown* (advertising brochure, San Antonio, 1913[?]).

newspaper before turning to spend the remaining forty years of his life growing pecans near Coleman in central Texas. The variety named after him became moderately popular and, when crossbred with the Mahan cultivar resulted in the protogynous Wichita, by far the most popular USDA cultivar. The Wichita accounts for 10.2 percent of the market nationwide. Although the tree itself has weak limb connections, which make it vulnerable to ice and wind damage, and requires perfect soil, space, climate, and care, these special demands are considered a small price to pay for the high productivity and quality. Halbert built a reputation by pioneering topworking efforts, fostered in part because, he confessed, "I could never have any success in my section with spring budding or grafting. The atmosphere seems to be too thirsty, and absorbs the water before the buds or scions can start to grow." His ideas on topworking were ruthless. "The entire top from six to twelve feet from the ground must be taken off. If land overflows, go above twelve feet if necessary to get above high water mark. Do not cut below six feet for fear the cattle may rub off the buds. Leave no limbs on the tree to divert and rob the shoots of the natural flow of sap. . . . The pecan is very tenacious to life, and there is no danger of killing it unless too old to be budded. If it is too old for that purpose, you had better apply the ax to the root of the tree and let some sapling it overshadows have the sunshine." Appropriately, albeit sadly, he died at the age of seventy-eight after falling from a tree.

J. H. Burkett was born in Tennessee on a farm that became the site of the Battle of Murfreesboro. His family moved westward to Texas and settled in Burnet County. At seventeen years old, after his father died, Burkett left the classroom with what he gauged was about a third-grade education. He bought a farm in nearby Callahan County in 1899, and his two sons, Omar and Joe, discovered a pecan tree on nearby Battle Creek from which he grafted the Burkett cultivar. In 1916, after procuring funds from the legislature to undertake a special pecan study, Burkett was hired by the Texas Department of Agriculture as chief of the Division of Edible Nuts. Later, he headed the pecan division for many years and wrote in gratitude: "The taxpayers of the state sent me to school and paid all of my traveling expenses for more than ten years, which gave me an opportunity to try and satisfy my innate curiosity for trying to determine the 'why' of any and all of the hidden secrets of nature regarding our native pecan. So I actually feel that I owe the citizenry much more than they owe me. Most all of the fun and pleasure was mine."

Burkett's career spanned years of intense experimentation that carried the pecan industry hundreds of miles westward from the native belt, into New Mexico and later to Arizona. Such work demanded a wholly new type of cultivar, one that did not need to be close to a natural water supply, could tolerate aridity and intense sunshine, and could thrive at a higher altitude. The number of varieties proliferated, both for the pioneering regions and within the natural range. In his annual departmental bulletin Burkett cautioned growers eager to try new cultivars to select those that had originated under conditions similar to those in their own land holdings. And with this advice he added a warning that conditions could vary over short distances if, for example, the soil type changed.

He began to distribute scions at his own expense from his own Burkett cultivar to different parts of the state in 1911, but admitted thirteen years later, "it has not been sufficiently tested for us to determine for certain just where it will succeed." Although not a front-runner, the Burkett is still ranked in national production. When Burkett solicited results for county agents and growers for his annual bulletin, he learned from Erath County that the Burkett pecan was considered "a healthy grower, free from scab . . . a moderate regular bearer of high class nuts"; in Palo Pinto County it was the "best nut and most prolific grower"; in Somerville County it "bears a fair to slim crop most of the time"; and according to a Grayson County admirer, it "sells well on account of size and quality at highest price; has been most consistent bearer." A grower in Travis County, well south of its original propagation, found it "the finest pecan I have found for quality but is more attacked by case-bearers, and generally bears less than Texas Prolific."

The Texas Prolific, one of Risien's cultivars, came in for much praise during the 1920s. In Erath County, just two steps northeast of its birthplace, it was considered "an early and productive variety of very pretty and delicious nuts," and in nearby Brown County it "scabs in the bottoms, but not in Brownwood."

Halbert's cultivar, from the same general area, drew mixed comments, as it had unresolved problems with infestation. Near home it was "very prolific, very profitable in upland, scabs in places with poor drainage," and "outsells all others on account of price." To the west, in Kimble County, one grower "had a nice crop of Halberts in 1924 but over one-half were infested with weevils," and farther south in Bexar County, near San Antonio, one respondent lamented, "I consider the Halbert a complete failure in my grove."

Several eastern varieties were well received in Texas, especially the Success from Mississippi. Along the river bottoms in Guadalupe County, there was some slight reservation, as one crop, despite being a good bearer with very fine nuts, was badly affected with an "obscure scale." But in Navasota, a riverine town in East Texas, the Success drew unqualified praise: "If I were to plant over I would plant only Success for following reasons: ripening early and uniformly. Free from scab or rosette. Large nut well filled and easy crackers. What more could you want?"

The mood in Texas encouraged a competitive edge, with some orchards set to both experimental and established cultivars that produced mixed results. Mrs. John Kemper in Grayson County tried to rank her preference among a dozen varieties and conceded that her report was "much jumbled and on several points indefinite. But with so many varieties and several too new to feel sure of, it seems about the best I can do at present."

Burkett and his staff faced a serious problem with rosette, a plant disease which appeared as "a dense cluster of poorly developed foliage in the form of frizzles or rosettes at the twig terminals." His field researchers in Winter Garden, an area in southern West Texas that covers part of Zavala and Dimmit counties, were hoping to discover a viable treatment. They found, fortuitously rather than intentionally, that when nutrients were applied from a zinc-coated bucket, the rosette symptoms began to disappear. The deficiency that caused the leaves to wither was simply an insufficient supply of zinc, for although the soil might have an adequate amount, a high pH might render it inaccessible to the roots of the trees and thus unavailable to the leaves. As the disease worsened, it halted nut productivity and weakened the trees. Now, it is standard procedure to apply zinc several times early in the season to prevent the trouble.

Among Burkett's staff, a young man named Louis Romberg took on the pecan division of the USDA research laboratory then headquartered at the University of Texas at Austin. Seedling trees were planted in a leased orchard there, which was inadequate as the work expanded, so in 1938 the equipment and personnel were moved to Brownwood, where the city supplied a sizable tract of alluvial soil for long-term use.

Romberg is popularly credited with the idea of naming the improved USDA varieties for Native American tribes. After the first USDA cultivar was named the Barton, for the man on whose land the propagated tree stood, all the others recognized the tribes of the United States, including the Apache, Caddo, Cherokee, Cheyenne, Chickasaw, Choctaw, Comanche, Houma, Kiowa, Mohawk, Oconee, Osage, Pawnee, Shawnee, Shoshoni, Sioux, Tejas, and Wichita.

As the pecan industry increases, the counties west of Interstate 35 continue to rank higher in the share they contribute to the total production. The 1920 U.S. census recorded eight trees producing a total of ten pounds in El Paso County. Eighty years later the figures had risen to 251,313 trees producing 7,526,620 pounds. In Hood County, near where Risien, Halbert, and Burkett carried out their work, the Brazos River nurtured a harvest of 343,946 pounds in 1920. This figure has increased tenfold in recent years.

Pecans in Texas, then, follow two paths. The state supplies the lion's share of seedling nuts nationwide, some twenty million pounds. Specialists agree the figure could be multiplied and reach as high as a hundred pounds an acre in a good year, especially if the orchards are managed more stringently and thinned out to allow good trees more growing room. If the present trend in environmental wariness continues, it is not inconceivable that many pesticides will be banned, which some believe will reduce the survival of those cultivars considered more chemically dependent. Should this happen, the native trees in Texas, together with those in Oklahoma and Louisiana, would be in great demand.

While this is not beyond possibility, the second path pecans have taken in Texas is in the development and widespread planting of western varieties. The cultivars, radically different from those in the East, have extended the growing area by several hundred miles, reaching out into the desert Southwest to the new, irrigated croplands where once the most visible tree was the mesquite. The combination of ingenuity and science, already seen in biological insect control, may lead to other ideas that are environmentally acceptable but do not limit cultivation.

# ▪ ORCHARDS

Pecan trees are either seedlings or cultivated varieties that have been transplanted at some early stage of their growth. The seedlings cluster along the river bottoms and stream banks or in managed groves. The cultivated trees grow in planted orchards. Pecan trees, like those who grow them, confront particular needs. For the trees, the needs are wholly physiological: they will thrive in places where the environment contributes to rather than detracts from their well-being. For the commercial grower the needs, almost by definition, are financial. They range from the selection of an orchard site through several aspects of management to harvesting and, finally, marketing.

Native seedlings may produce between 500 and 700 pounds an acre. They produce heavier crops when the groves or orchards where they grow are managed rather than left untended. After an orchard planted with improved varieties matures, it may be expected to yield an average of 1,200 pounds to the acre. There are tales of extraordinary productivity in single trees, but these numbers often increase in the telling, like the size of an angler's catch. Outstanding trees have certainly existed. In some instances their prolific habit generated prominent cultivars, and their annual crops have been carefully recorded, showing that an outstanding season was not merely a fluke. The original Halbert tree averaged 200 pounds each season over a ten-year period and one year produced 400 pounds. The San Saba, monitored over twenty-eight years, once yielded 800 pounds, but averaged 215 pounds. The first Oliver tree bore about 500 pounds a year for thirty years and on one occasion topped 800 pounds. A seedling near Bend, Texas, sounding more at home in wonderland, yielded 1,400 pounds one year and, so the story goes, if the hogs had not eaten so many the total might have reached a ton. In 1941 the American Pecan Journal and Southern Horticulturalist featured the Nogal de la Música (named for the melodious sound of the wind rustling through its leaves), a Goliath that grew 160 feet and had a limb spread of 150 feet, in southern Chihuahua, in a renowned apple-growing region some three hundred miles south of El Paso, Texas. The owner claimed an average annual yield of about one ton.

For pecans to thrive, beyond Burkett's comment that the constitutional vigor of individual trees is a factor, they need a suitable climate, adequate water supply, protection from predators and disease, and well-nourished soil. The nutrients may be applied directly to the soil with chemical or biological fertilizers, drawn from other plants covering the ground, or intercropped between the trees.

The appropriate climate for pecans is quite wide-ranging, as the native belt indicates. The weather must be warm enough to sustain an adequate growing season, with a chill spell during the winter dormancy period. An annual crop can be destroyed by extreme conditions such as the lengthy drought that killed millions of trees in West Texas in 1917. Hurricane Camille uprooted over 50 percent of the pecan trees in Mississippi in August 1969, and seven or eight million pounds of pecans perished. Hurricane Alicia caused massive destruction in 1983.

Pecan trees need one to two inches of water a week during the growing season, which occurs between April and October, or about fifty-five inches a year. Water is especially needed to initiate nut growth in the spring, in late May and June to increase the size, later in the summer to prevent the nuts' drying out and dropping prematurely, and prior to the shucks' splitting open in October. As rain falls unevenly and unpredictably, planted trees need supplementary irrigation. Flood irrigation is ideal for flat terrain. This practice involves wetting the entire soil mass, and as it is simply an extension of nature's way, the trees can absorb the water without problem. Where the terrain is rolling, an alternative method is necessary. Until recently the answer was a sprinkler system that was not only expensive to install, but also extravagant as so much of the water sprayed into the air was lost to evaporation. New technology offers a high-frequency trickle and microspray network that can be carefully regulated to provide each tree with an appropriate ration. The water travels along perforated pipes and is microsprayed over a prescribed area, perhaps four feet in diameter. The system works well whether the land is flat or rolling, and as the water is diffused under low pressure, there is little evaporation. It allows total monitoring, so it is cost effective and wins good grades from environmentalists who are concerned about wastefulness. Once the water quantity has been determined and made available, and assuming no salinity, the trees are unconcerned about whether their water arrives as rain, from a river, or from a well.

While having the water table within reach of the tap root is practical, the holding capacity of the soil is more important. In the old days, the rule of thumb to make planting pecans worthwhile was to have soil that could produce one-half to one bale of cotton to the acre, or from thirty to forty bushels of corn. Once an orchard site was selected and cleared of other trees, cultivating for a year before planting was recommended. If the land had already been cultivated, a few months under barnyard or green manure did the trick to supply the soil with nitrogen.

Today, while there are nitrogenous chemical fertilizers on the market, arrowleaf and crimson clover are still chosen in the southeastern states, as they can be counted on to fix about one hundred pounds of nitrogen to the acre. Beggarweed, velvet and soybeans, and hairy vetch are among other choices that are favored. As well as supplying nitrogen, legumes reseed themselves and are not hazardous to the nut harvest. Various cover crops that nourish the soil and prevent leaching can be grown between the trees to provide both short- and long-term benefits.

Early-maturing vegetables, which are ready for harvesting in May, can be planted for short-term advantage. Some growers recommend a three-year rotation that includes cotton, corn, and cowpeas. Over the long term, intercropping is generally an economic necessity, producing income while growers wait for the trees to mature. In Georgia pecans have been profitably intercropped with peach trees, and in Alabama and North Florida satsumas have shared space in orchards. Several types of fruit trees are short-lived, and as their productivity wanes, the pecans are ready to bear.

Another way of nourishing the soil is by utilizing the space as pasture. Munching cows provide a bucolic picture from the roadway, though growers have learned that if the cattle are not removed in time they can trample early falling nuts. Additionally, the manure affords a breeding place for aflatoxin, a mold organism that may afflict nuts lying on the ground.

Whichever way the soil is improved, the results are seen in the yield, which reflects long-term rather than yearly attention. Beside nitrogen featuring in the annual diet, potassium, calcium, phosphorus, and several trace elements must be supplied in alternate years. Even seedling groves, supposedly living as nature intended, can be helped. In the past they were frequently neglected, assumed as it was that nature would take care of matters and the seasonally flooded soil would supply all necessary nutrients. But the crops were seen to improve once the underbrush and other trees were removed. In prolific seasons, barren or shy-bearing pecan trees could be spotted, and they, too, were

In the early years of pecan cultivation, various means and solutions were tested to determine successful methods of protecting crops. *Photo courtesy USDA.*

taken out so the heavier-bearing ones had more space. Growers help by providing adequate soil and water and planting where the climate is friendly. They themselves have additional concerns. Above all, they need money, for even with human and technological innovation, it will be several years before the pecans generate an income. Growers must select a suitable orchard site and establish it after a number of interwoven details are in place. They must develop a sound management program to see them through each year's hazards and setbacks and enable them to harvest on schedule. Finally, they must find markets.

Funding takes priority. The larger the orchard, the more money needed, and the greater the need for a sound business plan. Despite topworking and cultivars suited to particular regions, seven to ten years will elapse before production becomes a reality. Intercropping is the likeliest buttress.

Once assured of enough funds to keep going, the next consideration is an orchard site. Money plays into this too, for the location may need intensive work before it is ready for planting. Direct cost is only one consideration. Growers must be sure they can hire seasonal labor and will have access to markets and processing facilities. Attention turns next to establishing the orchard. The land use should be planned on paper, following a choice of patterns.

Trees may be planted in a square (the most popular layout), rectangle, or triangle, or in diagonal rows. Sixty years ago growers believed fifty to ninety feet had to be left between one tree and the next, but fifty feet is now considered enough. As they grow, some trees can be removed to allow more room. In rectangular orchards pecans may be planted as close as thirty feet apart, leaving alternate rows sixty to ninety feet wide for intercropping. In the United States these rows will likely be filled with short-lived fruit trees, cotton, milo, vegetables, or wheat. In Mexico, grapes have been successfully intercropped.

It is also important to figure out which varieties to plant. Protandrous and protogynous varieties must be selected, to ensure pollination. Trees may be purchased at different stages of growth. Despite the higher cost, it is generally advisable to plant trees that are three or four years old and five to seven feet high, rather than planting the nuts themselves. Although the price of nuts is very low, it is harder and far costlier to control weeds so close to the ground.

Growers have to consider ways to control pecan diseases. Scab, prevalent in the Southeast but coming well into Texas, is among the most tiresome challenges. The fungus develops in tiny pustules and is spread by insects or the wind. The Stuart cultivar was immune to scab until about 1920, but ten years later the fungus was showing up and by the mid-1950s could be held responsible for measurable economic loss. Among several other diseases are those caused by mineral deficiencies: rosette, resulting from a shortage of zinc, and "mouse ear," from manganese deficiency.

With all the considerations and chores, growers must never lose sight of their goal: to harvest a bountiful crop of pecans each fall once the trees have matured. The harvesting season is short and intense, as the nuts are most in demand for the winter holiday season. As far as possible, preparation for the harvesting process has to be in place by August. Even if everything goes according to plan, there can still be upsets. Labor may be inadequate or unreliable, the weather can turn nasty, harvesting equipment can break down, animal predators may be particularly obnoxious, or the pecans may ripen late.

Harvesting can be done manually or with varying degrees of mechanization. In the past, the latter was not an option, and harvesting was enjoyable teamwork as well as heavy toil. At the Twenty-Sixth Annual Convention of the National Pecan Growers' Association, W. J. Millican described the Texas way of harvesting pecans: "The method is by first preparing a rope ladder like used by the sailors. . . . The length of the ladder is about 20 feet. This usually puts the climber to the first limb. . . . To the top of this ladder is tied a half inch rope, we call the fall line. That is put through the fork with a fish cord with a weight on the end to pitch through. . . . We are now ready to ascend. The man that threshes the tree has a small rope around his waist. He climbs to the top branches and lets the end down and his pole, about ten feet in length, is sent up to him. This pole is of mountain cedar which has been cut some 20 days, pealed [*sic*] and seasoned in the shade. It is tough, yet limber. The average day's work at this season is 250 to 300 pounds per day, some trees making from 200 to 1,000 pounds. The average pickers to each thresher is four. The picking is not started until the tree is threshed, thus preventing the danger of anyone getting hurt by falling dead limbs. The picking starts from one side, the pickers abreast, within arms' reach on each side. When this is over they start and cross pick, going over the same ground. The best picker gets the nuts left by the sorry picker. These are put in bags holding from 20 to 30 pounds. Each picker has his sack tied around his waist and when filled it is sewed up and left under the tree. . . . We find that women are the best hands for picking pecans. . . ."

Today, orchards which are harvested manually are usually smaller than twenty acres. The orchard floor is cleared, then the threshing begins, with rigid cane or PVC poles. The capital outlay is low, but those hired to help need close supervision as they tend to slow down after a while, cluster where the nuts are thickest, have difficulty reaching higher branches, and, on occasion, have been known to steal.

Mechanical harvesting starts when about 60 percent of the shucks are seen to be split. A mechanical shaker is set in motion, followed by a series of machines that have little need for human hands. Some growers now choose to install storage, processing, and packaging equipment so that they are ready to market directly from the orchard.

# ▪ ANIMAL PREDATORS

Halford A. Halbert, who grew pecans in Texas at the turn of the century, believed that "all animal life loves the pecan from the tiny ant to the giant elephant." Even if he exaggerated at the heavy end of the scale, squirrels, bluejays, and crows lead the field in causing astounding damage to pecan crops. In 1982, a bad crop year by any standard, animals were held responsible for a third of the loss in pecan yield. Their appetites could not be satisfied by the supply of food in the wilds, and they simply migrated to the orchards to feed. An Oklahoma study revealed that over one two-year period 19 to 30 percent of nut loss in the state came from wildlife. While insects tend to provoke fear and revulsion, most of us are conditioned to find small animals and birds winsome and, unless directly affected by the damage they do, are reluctant to see them as culprits.

For a few years a bumper sticker in the pecan belt read, "Eat pecans—millions of satisfied squirrels do!" Only outsiders found it amusing. The eastern fox squirrel (*Sciurus niger*), eastern gray squirrel (*Sciurus carolinensis*), and, in more limited territory, the southern flying squirrel (*Glaucomys volans*), devastate crops. They are among the few animals that damage pecans before they ripen, munching on spring nutlets in addition to mature nuts and in winter enjoying the wood of young trees and tearing into the bark.

Squirrels build their nests in late May, when unfurling leaves provide hiding space. In July shotguns blast some of their offspring out of ill-concealed nests. But squirrels are elusive, constantly migrating in search of new homes with better food supplies. They prefer to eat in the early morning, from about six to nine o'clock, and their idiosyncratic methods incriminate them as they leave broken nuts strewn around. The eastern fox squirrel leaves a jagged entrance hole in the pecans with a few large gnaw marks, while southern flying squirrels gnaw through the long side of the shell, leaving small, barely noticeable tooth marks. A squirrel can eat four ounces of nuts a week and bury another two pounds.

The first practical idea for coping with squirrels came from Edmond E. Risien, who fastened a circular sheet of tin midway up the trunk of the tree, at an oblique angle, to thwart squirrels trying to scramble up. Today's improvement is a spring trap on a platform nailed some four feet from the ground and baited year round with pecans, peanuts, or even peanut butter. Some orchard owners favor corn bait, but neither lure wins full marks. Exasperated cultivators also resort to twelve-gauge shotguns and number eight shot. They enlist the help of a good squirrel dog to bark frantically at the base of the tree, transfixing the squirrel, which then becomes an easy target. This method has a social dimension, gathering friends together with the promise of a goodly meal of squirrel meat. The ideal solution, not always feasible, is to plant new orchards in areas fairly free from wildlife, with the perimeter of the site at least a fourth of a mile from woods, and be prepared to harvest the pecans quickly when the time comes.

In Oklahoma the crow, which imperils the entire pecan range, is considered an even worse menace. Intelligent and cautious, the birds madden the growers, who hear threats to their livelihood in every different tone the crow can muster as it calls to other crows. Adding insult to considerable injury, *Corvus brachyrhynchos* selects the larger, best-tasting, thin-shelled nuts and prefers cultivars to natives.

As if the resident population were not enough, nesting in the orchards and producing four to five offspring, flocks of migrants come in fall and stay throughout the winter. The local birds act as hosts, showing the visitors the lay of the land. Crows spotted in the trees in early September are testing for ripeness, for nuts they can extricate from the shucks. As they feed, scouts protect them from any intruders wanting a share, and flocks have been known to kill or run off a scout bird for unsatisfactory performance.

Like the squirrels, the crows are not casual nibblers. Each can destroy a pound a day from September till harvesting is completed, hammering at the center of the nut and smashing it into several pieces. Usually their appetite fails to match their enthusiasm, and ants finish up the leftovers. All the same, they can eat an ounce of nut meat a day, amounting to about 6 percent of their body weight, then carry off the rest for later.

As they glut themselves, they drop pieces and dart down to the ground. They are caught when they eat the grains of yellow corn scattered at the foot of the trees as bait and are driven away by propane scare guns. This method is rather less effective, if more humane, than the old method of dynamiting them by the thousands as they roosted. A hunter's bounty, which is considered at least token compensation, is paid after November 1.

The crows' relatives the blue jays (*Cyanocitta cristata*) are not as intelligent but hold the honor of being the worst predators in native orchards, choosing the smaller pecans with pointed ends. They shrug off scare devices and chemical bait, and it is uneconomical to shoot them. Migrating flocks join the resident community in late September, but they are weak flyers and daunted by long open spaces. They prefer to follow a tree line to the orchards and, fortuitously, enjoy blackjack acorns more than pecans. It is expedient to surround the orchard with oak trees if this is an option.

While these two species are responsible for most of the

damage, other birds known to break pecan shells and eat the meat are the rose-breasted grosbeak, northern blue jay, white-breasted nuthatch, yellow-bellied sapsucker, and red-bellied sapsucker. Not only squirrels will make a meal of the nuts and bark, but also black bear, gray fox, cottontails, raccoons, and in the North the fox squirrel and red squirrel. There are several smaller animals partial to nuts and leaves, among them chipmunks, pocket mice, the white-footed and prairie mice of Oklahoma, and both eastern wood rats and cotton rats. Deer can be destructive, and cattle, which usefully crop the grass, must be removed in the summer before they heavily trample the ripening nuts as they fall to the orchard floor.

When Edmond Risien worked among his pecan groves by the San Saba River a century ago, he perceived a menacing change in the wildlife population around him. The wild turkey was becoming extinct, and there were notably fewer insect-eating wild birds. He knew that in their absence, insect pests would multiply and could destroy not only his work with new cultivars, but also the luxuriant groves of seedling pecans. Dismayed, but not distraught, he came up with a gentlemanly resolution—domestic turkeys and guinea fowl should be invited into the orchards to satisfy their appetites. This option was revived with a sophisticated slant generations later, when the damaging effects of wonder chemicals were recognized. A guest list was drawn up, naming the beneficial insects such as ladybugs, lacewings, and praying mantises that would forage among pecan trees for victims such as aphids to satisfy their appetites.

About twenty species of insects and mites damage pecan trees, attacking the nuts, leaves, stems, buds, root, or bark. Their degree of success varies according to the geographic area, pecan types, and vagaries of the weather. In the past, pest management was generally handled culturally. The Chinese used fire to sanitize cropland and reduce pest problems, and over two thousand years ago controlled citrus pests with the help of ants. In the nineteenth century inorganic pesticides like lime sulfur heralded the age of scientific control. Numerous "improvements" followed. When DDT was reintroduced during World War II, it was an apparently ideal way to rid soldiers of body lice, which spread typhus, and mosquitoes that carried malaria. The miraculous short-term success led to the development of other synthetic pesticides, such as chlorinated hydrocarbons, organophosphates, and carbamates that also killed the pests responsible for crop disease and destruction. When the long-term negative effects were finally recognized, other means of combating pests had to be found.

In recent years the rallying cry has become IPM—Integrated Pest Management. Crop protection programs have been tailored to reduce chemical use and develop alternate strategies. Several details have had to be addressed, such as the early detection of pests, an assessment of their density, changes that may be expected, the effects of the various control options, ways of implementing control, and costs. To bring all this together, scientists and cultivators have had to be familiar with the life cycles of the various pests, the pros and cons of available insecticides, characteristics of the targeted crops, and the chance of unpredictable weather.

The destructive power of one insect, the pecan nut casebearer (*Acrobasis nuxvorella Neunzig*), may account for the loss of 25 percent of the pecan harvest in Texas. West of the Mississippi, and reaching as far as El Paso, the nut casebearer does more damage to pecans than all other pests combined. The nocturnal adult moth is about the same color as the bark of the tree and also very secretive. The female lays 50 to 150 eggs in the grooves that slope into the base or tip of the pecan nuts when they reach pea size in the summer. In less than a week the eggs hatch. The larvae do their teething, so to speak, on the buds, then clamber back up to feed on the nuts until they are full grown. When the supply of pecans is good, their 42- to 45-day life cycle may be repeated for four generations in a season.

East of the Mississippi pecan leaf casebearers (*Acrobasis juglandis*) winter as larvae in cocoons attached to the tree buds and smaller branches. When the buds swell, the semi-dormant larvae emerge to feast ravenously on the new leaves. They pupate in late spring, and the moths lay eggs on the lower surface of the leaves along a vein or midrib. The young larvae, which hatch six to nine days later, eat the food on hand. On maturation they transform into encased pupae, and when the summer ends abandon these cases and migrate to the buds to spin hibernacula and start the cycle again.

Pecan weevils (*Curculio caryae*) destroy nuts in two stages of their lengthy lives. A small brown beetle tinged with gray, the female will puncture ripening nuts with her long snout and place her eggs in the kernel. Once hatched, the larvae will demolish the nutmeat, then cut a circular hole and crawl out—rather like a comic-book crook departing jail—and down the tree to winter several inches underground. They may elect to stay in the earth for two winters, pupating there in the fall. Although the metamorphosis occurs in the space of three weeks, the adults lie low until the summer, then fly up to deposit their eggs and trigger a new generation. Apparently the adult weevils return to the tree in which they fed as grubs, if nuts are available there, so that certain trees may be heavily infested year after year. Late-maturing cultivars such as Success and Van Deman usually escape attention if the weevils have found enough nuts to supply both food and egg-laying sites in neighboring earlier-maturing cultivars.

These infesters, along with spittle bugs, the southern green stinkbug, black and yellow aphids, and several others are removed by a calculated spraying schedule and, increasingly, insects like ladybugs and lacewings, which prey on aphids, and trichogramma, a species of parasitic wasps that lay eggs inside the eggs of moths and butterflies, among them pecan destroyers. Trichogramma eggs may be shipped already inside the host eggs, and lightly glued to ten-by-three-inch cards holding 125,000 eggs each. Once the eggs reach their destination, growers are ready to hatch them as their pest control program demands, and the almost microscopic wasps emerge after eight to ten days. They are then released amid the pecans, either by hand or, if acreage warrants, by air, ready to spread their tiny wings and set to work, their offspring eating the pest eggs before the latter have a chance to hatch.

Small mammals and birds—to put in a good word—have expanded and enriched the pecan range by dropping or burying countless nuts.

# THE PECAN INDUSTRY

In his book *The Pecan and Its Culture*, published in 1906, H. Harold Hume pondered a question still on people's lips today: What is the ideal pecan? "Ideal pecans? Yours or mine or your neighbor's? Whose? Ideal as a dessert nut or for commercial purposes? Which? Ideal for Texas or Alabama or Georgia? Where? Ideal early, medium or late? When? Two inches long or three and a half? Don't you see that there may be as many ideals as conditions to be filled, and that ideals exist mostly in men's minds?"

Generally to the consumer, bigger is better, "even if as large as my pecan was reported to be when it took the premium at Waco in 1901—viz., the size of a hen egg," wrote grower Halford Halbert with pride, but at the same time he declared his own ideal to be "one weighing sixty to the pound, round in shape, the shell a mere film, which can be easily cracked in the palm of the hand, kernel easily cleaned and of sweetest flavor, nut of brown color with dashes of black at the apex."

When members of the National Nut Growers' Association came up with a tentative rating system for pecans at their second annual meeting in 1903, their allocation of points was less remarkable than the fact that such a rating could be considered. Until this time quantity had been the sellers' main consideration. But change was on the way. At the Texas Nut Growers' Association convention held in College Station in July 1907, the British-born naturalist H. P. Attwater reported: "A certain amount of rivalry among the hundreds of white people, Mexicans and negroes who make up the great host of those annually engaged in the industry of nut gathering, stimulated by the receipt of a better price for pecans of extra size, has undoubtedly been the cause of active search and discovery of most of our finest pecans, and while the whereabouts of some unusually fine trees has been kept a secret by the finder for a certain length of time, sooner or later the localities become known to others."

His comments came as pecans were becoming a major commercial venture. A first dramatic shift had occurred when pecans, bartered between Indians and newcomers, became a cash crop during the nineteenth century. Ferdinand von Roemer, the German botanist who traveled through Texas in 1848 observing the fauna, noted: "Everywhere in Texas, especially in river bottoms grow pecan trees (*Carya olivaeformis*) which bear fruit resembling walnuts. Whole wagon loads of these are gathered in the fall and brought to Houston, from which point they are shipped to the Northern States."

Indians, too, began to prefer cash rather than goods for the pecans they gathered and clearly had long recognized their worth. Article 11 of the Crimes and Misdemeanors of the Laws of the Muskogee Nation, 1890, stipulated that "if any person cut down or destroyed a pecan tree for the purpose of obtaining nuts the fine was twenty-five dollars for each tree." If the suspected party denied the offense, then the Light Horse Company, the body of men serving the Five Civilized Tribes as a mounted police force, was obliged to carry the suspect before the district judge for trial.

Lucy Gage, who taught a generation of elementary education majors at Peabody College in Nashville, remembered her childhood on the Kiowa-Apache Reservation in Oklahoma and the time when she and her companion "came upon some young Kiowan Indians with a wagon load of pecans from the southern part of the Territory. We stopped to buy some but first wished to

San Antonio was the hub of the Texas pecan industry for many years. *Photo courtesy USDA.*

sample them. Never shall I forget those blanket Indians vieing with each other in cracking nuts with their strong white teeth. They really made a game of feeding the two white women pecans until we made our purchase."

As early as 1802 the French were shipping Louisiana pecans through New Orleans to the West Indies, and Houston became the point of dispatch for Texas. Settlers in the central part of Texas also viewed pecans as a cash crop. San Antonio became the regional hub and in 1880 harvested 1,250,000 pounds. The sellers received five to six cents a pound, and H. P. Stuckey, long-time director of Georgia's Agricultural Experiment Station, knew, like Attwater, that but for this trade, "the people of some localities must have starved for lack of remunerative labor. Hundreds of both white and colored people go out with horses and wagons to gather those nuts."

Agricultural statistics in the thirteenth census of the United States, taken in 1910, showed overall pecan production had trebled in a decade to 9,891,000 pounds, valued at $972,000 and trailing only the 22,027,000 pounds of English or Persian walnuts valued at $2,297,000 and the numerous but less prized harvest of black walnuts, which reached 15,629,000 pounds.

While the railroad snaked across the South and Southeast, substantially increasing the quantity of pecans transported to coastal ports after the Civil War, the most radical changes were brought about a generation later by three scientific advances. As these advances converged, they transformed a locally profitable business, which was little more than a sideline, to one of the most economically viable native crops in the United States.

The first development was botanical: genetic manipulation introduced varieties adapted to conditions that existed beyond the native pecan belt. Today 700,000 acres of native pecans are still harvested in Texas, but on average these trees produce only twenty-eight pounds per acre. In comparison, some cultivated varieties produce an average two thousand pounds an acre. The expansion of territorial range is even more remarkable, resulting from painstaking research that allows cultivars to thrive in climatic and soil conditions alien to the natural habitat. In Georgia, for example, far from the nurturing Mississippi River, a 500-tree orchard was planted in 1887 near Albany, by 1910 some 450,000 pecan trees had been planted across the state, and ten years later a million trees produced 2.5 million pounds of nuts. Enthusiasm for the miracle crop knew no bounds, and between 1910 and 1920 northerners contemplating retirement were lured to Georgia by the promise of a business that virtually ran itself. One company sold 25,000 acres of orchards in five- and ten-acre units! Many smallholders were stung by this effrontery and forfeited their savings; their orchards, mostly bearing Stuart pecans, were infested by pests and disease. But others prospered, and the enterprise contributed to Georgia's status as the leading pecan-producing state.

The second improvement, more reliable storage, was brought

about by efforts to improve the overall quality of pecans. Specialists in horticulture, botany, nutrition, chemistry, food processing, and food technology shared what they learned in all aspects of nut production, so that while no single field of science is responsible for all improvements, several play a role. Thus scientists were able to determine the correct balance of moisture content, chemicals, and temperature to ensure the pecans a longer, sweeter shelf life. Spoilage had challenged growers and sellers alike since Jefferson wrote from Paris for his pecans to be shipped in a box of sand. Until science took on the task, the most satisfactory alternative had been sphagnum moss.

Although varieties differ from one to another, among seedlings as well as cultivars, generally speaking pecans begin to go stale and rancid if left at room temperature for more than three weeks, or at 10°C (50°F) longer than three months. Thus it has been important to come to grips with the matter of storing an ever-increasing harvest. For Harold Hume, storage was a key concern:

"They have found and are holding a place at the soda-fountain. Foods are manufactured from them, and they have become a source of fat for the vegetarian. . . . Unfortunately, the stock is not kept in such a way as to create a desire for more on the part of the consumer after he has tried one package. They are frequently old, stale, and rancid. When the kernels are carried through the heat of the summer in an ordinary jar" the results were not palatable.

Since he drew attention to this problem, there have been years of experimentation, and now we know that in order to prevent mold, discoloration, the breakdown of oil, and the shells' sticking too tightly, pecans need to be dried to 4.5 percent moisture content as soon as possible after harvesting. Moisture control is the top priority during the harvesting, storing, and processing periods. Although reduced during the storage period, the moisture level must be elevated for shelling to prevent the now brittle meat from shattering. The most common method to bring the moisture content up to an optimal 8 percent is by soaking for twenty to thirty minutes in cold water containing one thousand parts per million of measured chlorine, then draining and holding the nuts in sacks, vats, or barrels for sixteen to twenty-four hours, cracking them at any time during this period. Alternatively, they may be put under five pounds of steam pressure for six to eight minutes, then cooled and held for thirty to sixty minutes. This method is faster but runs risk of encouraging mold. With a third option, pecans are immersed for twenty minutes in water heated to 62.8°C, or 145°F, and shelled immediately after. In all instances the water enters the nut through the vascular system at the base, then travels along the middle partition to the apex.

Some pecans are refrigerated straight from the orchard, once the "pops," or failed nuts, have been removed along with other debris from the orchard floor. Until about 1960, ammonia was used as a cold-storage refrigerant for pecans, but if shelled kernels became exposed to ammonia gas, their appearance changed dramatically within minutes, leaving them black and unsightly. Even though the flavor and nutritional value were unimpaired, they looked so unappetizing they were unmarketable. Now, ammonia has been replaced by Freon as the refrigerating agent. Temperature remains an important concern and must be held at or below 48°F to prevent insect infestation.

The third innovation in the pecan industry was mechanized shelling. Gustave Antonio Duerler, a pioneer candy and cracker manufacturer in San Antonio, launched the pecan-shelling industry in June 1882 when he sent a speculative dispatch of fifty barrels of nuts, gathered and shelled by "friendly Indians," to cities in the East. Subsequently he hired local Mexicans, whom he paid to crack the nuts with railroad spikes, then pick out the kernels with tow-sack needles. A Swiss by birth, Duerler eagerly invested in Robert A. Woodson's 1889 invention, a cracking machine operated with one hand and fed with the other, and his subsequent power-driven, automatic, self-feeding machine in 1914. Just four years later, in 1918, Lee J. Meyer's more sophisticated automatic cracker opened the door to considerable expansion in the industry.

By 1928 the Duerler family, assisted by hundreds of young Mexican women, were shelling about one thousand pounds a day. The R. E. Funsten Company of St. Louis was responsible for most of the northern shelling.

In 1926 Julius Seligmann, a merchant and landowner in the small town of Seguin, Texas, where pecans cascaded down along the banks of the Guadalupe River, invested fifty thousand dollars to organize the Southern Pecan Shelling Company with a partner, Joe Freeman. Ten years later their business grossed three million dollars. They had forsaken mechanical equipment and employed shellers and pickers to work by hand, using old buildings and a minimum of machinery. The workers needed a considerable amount of space, and a system of contract labor developed. Contracting agents were obliged to buy the pecans from Seligmann's company, then sell back the shelled nuts. The company profited by about forty to sixty cents on each one hundred pounds shelled. By the mid-1930s San Antonio had some four hundred shelling establishments with forty to fifty workers in each. In 1935 there was a bumper crop, 105,000,000 pounds. Southern Pecan bought up the surplus pecans at three to six cents a pound and stored them in its adequate facilities. When the next harvest reached only 40,000,000 pounds, following the trees' natural pattern of alternate bearing, Southern Pecan made a fortune selling the surplus held in storage to smaller shelling-plant operators.

Despite the amount of money generated, the industry was unpopular among the workers and was poorly organized, but San

Antonio remained the center for two reasons. First, there was a superabundance of pecan trees in the area, and second, there were many "Mexicans" anxious for work. Although termed Mexican, many of the workers employed in pecan-shelling operations had been in the United States, often as citizens, for many years. The first sizable wave of Mexican immigrants came north across the border after World War I, displaced by domestic upheavals and tempted by the labor shortage. In the 1920s, once Asian immigration was halted, the demand for cheap labor increased in both cotton and agriculture. San Antonio attracted immigrants from the South, as the predominant community in the city was Mexican American, especially in a four-square-mile area on the west side, providing seasonal work within easy reach. The "Mexican" population in San Antonio tripled between 1910 and 1930 and then continued to hold steady. Although the pay amounted to only five to six cents an hour and working conditions were extremely poor, during the Depression years some twenty thousand people found employment in the shelling plants, although the jobs, lasting from late November through March, were taken as a last resort. The men often moved from place to place during the day seeking the best supply of nuts. The women put in a few hours of work while their children were in school, hurrying home to fix lunch and dinner. Even the children themselves, the elderly, and the disabled participated, despite grim and crowded conditions where many suspected that the brown dust hanging in the air was responsible for the prevalence of tuberculosis in the community.

As the Southern Pecan Shelling Company resisted automation and was totally reliant on hand labor, so other companies followed its lead. Three serious efforts were made to improve the situation. In 1933–1934 the National Recovery Administration attempted to stabilize the industry and raise wages. This initiative failed because of a dispute between the northern shellers headed

Pecans were often sold through family businesses before improved storage and the practice of mechanical shelling made pecans a highly profitable industry. *Photo courtesy USDA.*

by the R. E. Funsten Company in St. Louis and the southern ones represented by Southern Pecan Shelling. Several unions struggled to organize workers to help them seek better pay. In January 1938 Southern Pecan Shelling said it would lower the pickers' wages from seven to eight cents an hour to five to six cents. The announcement resulted in a spontaneous walkout, with union support from UCAPAWA, the United Cannery, Agricultural, Packing, and Allied Workers of America. Union membership swelled, and subsequently the shellers returned to work, accepting six and a half to seven and a half cents an hour.

The year's troubles were far from over. The Fair Labor Standards Act was scheduled to go into effect that October, setting a twenty-five-cents-an-hour minimum pay rate. With months to adapt to this prospect, San Antonio shelling companies chose to speed up production during the summer, paying five to eight cents an hour to workers. Southern Pecan Shelling shrugged off the act, saying pecan shellers would be exempt from the law as they were processors of agricultural products within the area of production. When this claim was rejected, the company applied for temporary "learners' exemption." This, too, was refused. On October 24, 1938, the Fair Labor Standards Act became effective, and the San Antonio operators shut down.

Several thousand shellers were thrown out of work, and the succession of events changed the face of the industry. By the following summer about half of the sixteen hundred shellers at work in the city were receiving twenty-five cents an hour, the rate established by the act. The workers who had shelled and picked so dexterously by hand soon mastered the new machinery. Gradually the industry stabilized and with the benefit of cold-storage improvements provided year-round jobs.

Today's automated equipment has been refined so that pecans may be oriented, then cracked individually by a plunger thrusting in about an eighth of an inch, cracking the shells longitudinally into about eight "barrel stave" pieces, each of which breaks in the middle. The shell is shattered into sixteen pieces without crushing or breaking the meats. Plants now boast batteries of cracking machines, each adjusted to specific sizes and able to handle quantities at a speed undreamed of by Seligmann and his contemporaries.

# ▪ NUTRITION

Many aspects of our behavior have become increasingly analyzed and formalized in the past few generations, including our eating habits, and recognition of the pecan's food use and value has changed. In the past the pecan was accepted as an important survival item during the lean months of the year and needed no explanation. Now its caloric, vitamin, and mineral properties have been determined and have passed muster, and it accompanies astronauts to the moon. The pecans that rounded off family meals, providing a chance to bond generations while nutcrackers and kernels passed from hand to hand, are today mostly marketed as shelled kernels.

But however and wherever they have been eaten, pecans have been enjoyed. Time and again they are remembered nostalgically, a reward for endurance or an unexpected treat. In 1854, when twenty-year-old Emma Altgelt traveled by wagon with her husband from the Texas coastal port of Indianola to settle and establish his law practice in San Antonio, the storms and loneliness suffered were forgotten briefly when she wrote: "As we passed over the bridge near Victoria and drove along the banks, everyone was amazed at the masses of nuts, pecans, which the norther had thrashed and we picked them up to our heart's content."

The pleasure persists and people still drive, in station wagons rather than the horse-drawn kind, simply to gather pecans for their own munching. As the nut crops grew and reached markets well beyond the native pecan belt, writers warned delicately of spoiling the pleasure by overindulgence: "They are likely to cause some discomfort or even distress, as would any other highly constructed food following a superabundant meal . . . particles of kernels not masticated are likely to pass through the alimentary canal as foreign bodies and not digest at all."

One pecan cookbook warned: "When pecan kernels are eaten raw they should be masticated and insalivated until reduced to a smooth paste, to insure the prompt admixture and action of the digestive fluids, and render them readily digestible."

Besides digestive risk, there were health dangers in the early years of pecan retailing. It was one thing to pick the nuts from surrounding groves and along the river bottoms, quite another to buy them. Customers were alerted that shopkeepers in pecan growing areas received their supply of shelled pecans for the trade from local customers who shelled them in their homes. Merchants were not questioned by their customers as to the observance of health regulations governing the preparation of this food product.

The Indians whom Cabeza de Vaca met survived two months of the year on nuts and "small grain," but neither their marketing nor digestive concerns were noted. In the early years of this century, people in rural Missouri, which lay in the native pecan range, fulfilled their dietary needs with pecans and other nuts and even included them as trading items on farms and plantations. But the practice did not prevail to the eastern edge of the belt. As late as 1919 the author of a history of Washington County, Mississippi, down in the delta region, noted with surprise: "These nuts are esteemed a very valuable food for swine, but since they have come into such favor as food for people they are now usually raised for the market."

As well as swine, pecans continue to nourish birds, squirrels and other small mammals, and even deer and cattle quite indiscriminately. For people, though, their popularity as a food—

rather than as a treat or fleeting escape from dieting—is dictated by the vagaries of fashion. In the United States, where hunger is a result of poverty rather than lack of available food, the fact that a diet composed of fruits and nuts contains sufficient nutrients in the right proportion to support life or that a pound of shelled pecans furnishes over three thousand calories is not especially appealing. Current concerns for appropriate body weight and a well-balanced, healthful diet mean the nut's high fat content is less welcome. Where once it was a boast to claim that half a Schley pecan contained sufficient oil to support a flame for eleven minutes, for many this is now a reason to avoid the nuts.

Although the flavor of pecans is tied to the oil content, there is no correlation between the quality of oil content, flavor, color, aroma, or texture and the size of nuts. But other determinants have been scrutinized, several by Texas students with a ready-made research topic on their doorsteps. In the 1930s, University of Texas student Willie May Wolfe Glover tested an early concern about the adequacy of Vitamin A content in different cultivars with the help of albino rats that endured an exclusive diet of pecans until their systems expired. The teams of rats that were fed Burkett and Western Schley varieties fared better than those limited to the John Garner and Mahan.

Genie Evans Cameron, studying at the Texas State College for Women, chose pecans as her research topic for a master's thesis in 1938, addressing their potential as a year-round staple food commodity. Her grass-roots approach and unequivocal belief that women had more time and a better grasp of food marketing led her to interview housewives to determine consumer demand and even to include in her thesis her dormitory's monthly food shopping list and menus. Her findings also demonstrated less controversial pecan benefits. Pecans are low in sodium, are high in protein and unsaturated fats, have no cholesterol, and are a good source of calcium, iron, phosphorus, potassium, and magnesium.

They compare well with other highly nutritious foods.

| *Material* | *Calories (per pound)* | *Protein Calories* | *Fat Calories* |
|---|---|---|---|
| Pecans (shelled) | 3,390 | 174 | 2,879 |
| Cheese | 1,888 | 470 | 1,376 |
| Beef loin | 1,160 | 335 | 825 |

A nutritional analysis shows that pecans, per hundred grams, supply:

| *Percentage* | | *Milligrams* | | *Milligrams* | |
|---|---|---|---|---|---|
| Protein | 9.25 | Calcium | 73.0 | Sodium | Trace |
| Fat | 71.20 | Phosphorus | 289.0 | Thiamine ($B_1$) | 0.86 |
| Carbohydrates | 14.60 | Iron | 2.4 | Riboflavin ($B_2$) | 0.13 |
| Fiber | 2.30 | Potassium | 603.0 | Niacin ($B_3$) | 0.9 |
| Water | 3.40 | Magnesium | 142.0 | Vitamin C | 2.0 |
| | | | | Vitamin A | (130 I.U.) |

Well over 80 percent of each annual harvest is now sold already shelled. Bakers buy some 25 percent of the shelled market, either as pieces or halves, and some also use pecan meal as a partial substitute for wheat flour. The pieces are available in different sizes and used especially in cookies, cakes, muffins, and breads. Halves, both large and small, find their way into pies and fruitcakes, where they also meet decorative requirements. Confectionery manufacturers buy a sizable percentage of shelled pecans for candy and snack items. Some are sugar-coated and resold, this time to ice cream manufacturers for praline ice cream, which ranks third in national popularity after vanilla and chocolate.

Pecans, like other foods, increasingly have their nutritional properties scrutinized. Some years ago food technologists sought a fresh alternative to freeze-dried meals for the astronauts participating in the Apollo 13 and Apollo 14 missions. The food had to be concentrated, high in energy, digestible, able to withstand temperatures far above and far below freezing, and, not least, tasty. These requirements ruled out apples, oranges, and other fruit that might otherwise have been welcomed, but opened the way for pecans. Their 3.5 percent moisture content, three-thousand-plus calories per pound, and 95 percent unsaturated fat composition provided an encouraging start. Added to this, a temperature of –170°F had done no perceptible harm, and raising the heat to a point just short of ignition merely roasted them.

The Desirable cultivar, weighing around forty to the pound, met NASA's stipulations. The nuts were first washed at a temperature nearing 150°F, then immersed in a two-thousand-parts-per-million chlorine solution. In order to ensure unbroken half kernels, an Inertia Nut Cracker was used, carefully cleaned in a detergent water solution, then in a two-hundred-parts-per-million chlorine solution. Once shelled, the pecans were vacuum packed in thick polyethylene packages. Thus pecans, which for thousands of years had lain on forest floors available to all who stooped to claim them, became the first fresh food to meet the stringent demands for a nutritional outer-space snack food.

# ▪ RECIPES

Nothing beats a good, fresh pecan cracked in the palm of your hand and chewed on the spot. However, this has not stopped growers and gourmets from sharing innumerable recipes in which the pecan is either central, an important addition, or merely a decoration. From the time American Indians pulverized the nuts to add to their staple bread or to ferment for greater pleasure, pecans have shown their versatility. Most pecan recipes fall into three categories: candies, pies, and fruitcake—although almost no one in the history of pecans, perhaps the world, has ever admitted to actually liking fruitcake.

One of the tourist attractions highlighted in a late-nineteenth-century guide to New Orleans was the meat market in the French Market where "Negro women station themselves . . . offering for sale 'pralines,' sugar cake made of pecan or peanuts . . . the location is near where Choctaws from north of Lake Pontchartrain sell herbs and medicinal plants."

Pralines were sold also in several shops on the 700 block of Royal Street that "specialize in the celebrated candy of old New Orleans—the praline. The praline is . . . a confection as old as the town itself. It is a crisp candy made from nut-kernels boiled in sugar. In France the praline was made by cooking almonds in sugar, called *almonde rissolée dans du sucre*. When the early colonists came to Louisiana the housewife substituted the native pecans for the almonds in making the candy."

Nowadays commercially made pralines are as popular as

From *A Book of Famous Old New Orleans Recipes Used in the South for More than 200 Years* (New Orleans: Peerless Printing Co. [1900 or 1901]).

those made from traditional home recipes. They are popular with the manufacturers, too, as they are simple to make on the production line and can be made in many sizes and shapes to sell at different outlets. Their quality remains stable under a wide range of temperature and humidity conditions, and they are easy to package, carry, and eat.

## PECANS

Praline recipes fall into two groups, the great divide being between the eastern chewy treat, with the addition of cream or butter, and the Texan/Mexican crisp, sugary confection. They require few ingredients, and the candy thermometers called for in many recipes were invented long after the first praline was savored.

One praline recipe contending for the "most basic" award was doing the rounds in the 1920s:

### Pralines (1)

Take 1 pound of brown sugar, 1/2 pound kernels, a tablespoon water. Cook sugar and water, stirring constantly, until it spins threads when poured from spoon. Then pour over little piles or heaps of pecan kernels arranged over a greased marble slab.

The equivalent southeastern-style recipe is:

### Pralines (2)

2 cups brown sugar
1/4 cup water
1/3 cup butter
2 cups chopped pecans

Stir water, sugar, and butter over slow fire till the sugar is dissolved. Add pecans; boil till mixture forms hard ball in cold water. Drop on waxed paper, allowing to spread 1/3 inch thick and 4 inches in diameter.

■ ■ ■

Cooking was an uncomfortable chore in the torrid South, and one early cookbook author recommended a quick nip from the bottle to make matters a little more tolerable:

## Pralines (3)

| | |
|---|---|
| 1 cup pecans | 1 cup cream |
| 2 cups brown sugar | Enough water to dissolve |

Cook till it strings. Stir in nuts. Add a touch of sherry or brandy.

## Pralines (4)

This praline never failed and allowed many girls to earn a livelihood making and selling them. The author claimed it was "the orthodox recipe of New Orleans."

| | |
|---|---|
| 1 pound light brown sugar | 1 tablespoon butter |
| 1/2 pound shelled pecans | 4 tablespoons water |

Set the sugar to cook; as it begins to boil add the pecans. Use just the halves as it makes a prettier piece of candy. Boil until the mixture begins to bubble and spins a thread, then take from the fire and beat well. Take a large kitchen spoon and dip it full of the syrup and the pecans and place each spoonful on a buttered marble slab.

■ ■ ■

In Texas pralines must be crunchy and sweet enough to set the teeth on edge. Raw Mexican sugar, piloncillo, generally sold in cones, has a distinctive flavor:

## Pralines (5)

| | |
|---|---|
| One 7-ounce piloncillo | 1/2 cup water |
| 1 teaspoon Mexican vanilla | 1 cup pecan pieces |

Dissolve the piloncillo in water over medium heat in a covered pan, breaking up the sugar as it dissolves. Take off the lid and continue cooking until the syrup reaches 238°F. Remove from heat and add vanilla. Add pecans, and beat until pecans begin to separate, then drop by spoonfuls onto wax paper. The pralines will crystallize as they cool.

## Pralines (6)

This traditional Mexican recipe bridges the divide between pralines and other pecan candies.

- 2 cups half-and-half
- 2 tablespoons vanilla
- 2 cups packed light brown sugar
- 2 tablespoons dark rum
- 1/4 cup light corn syrup
- 2 cups (1/2 pound) pecans

In a pan, combine the half-and-half, sugar, corn syrup, and rum. Stir until melted, and bring to a boil. Stir in pecans, and vanilla should be added here. Boil to 260°F, and drop onto a lightly greased jelly-roll sheet.

■ ■ ■

Once people can rely on an adequate food supply being available at all times, it is not long before they feel compelled to watch their diets. Statistics that separate necessities from options and reduce food to nutritional components earn respect when they reach acceptable conclusions. Learning from doctors and chemists that chocolate may, indeed, help counter depression is comforting. Though not quite as scientifically proven, candy improves bad days that are worsened by stringent diets. When a presenter at the National Confectioners Association conference told the audience that the "body" of candy is texture and the "soul" is flavor, this gave just the right touch of dignity to basic craving. However urgent the craving, people still have their favorites.

One cookbook writer and her friend cherished what each believed was her own grandmother's recipe. Later, they found it had appeared in an Oklahoma newspaper fifty years earlier, and everyone claimed to be related to "Aunt Bill" because of her marvelous candy:

## Pralines (7)

6 cups white sugar
2 cups cream or half-and-half
1/4 pound butter
1/4 teaspoon soda
1 teaspoon vanilla
2 cups pecans

In a cast-iron skillet place 2 cups of sugar and stir over a low heat, until the sugar begins to melt and caramelize. When it has liquefied completely, remove from heat. At the same time, cook 4 cups of sugar and 2 cups cream in a stainless steel pan. Stream the caramelized sugar into the boiling cream and sugar. Bring to the soft-ball stage, 236°F, then immediately remove from heat and set down the pan. Add soda, stirring all the time. Add butter and vanilla. Cool to lukewarm. Beat with a wooden spoon until mixture becomes thick and dull. Add pecans and pour quickly into buttered dish or onto wax paper. Cut into squares when cool.

## Pecan Logs

The Stuckey roadside pecan shops and restaurants originated in Georgia. Over the years their leading claim to fame has been a confection known as a pecan log. It makes long drives less arduous. This home recipe for pecan logs has been handed down for generations:

1 cup brown sugar
1 cup cream
1/2 cup sugar
3 tablespoons butter
3/4 cup maple syrup
corn syrup
chopped pecans

Cover and cook sugar, maple syrup, cream, and butter for 5 minutes. Uncover and cook to soft-ball stage. Cool for 15 minutes, then beat and roll in logs 1 inch thick and 3 inches long. Roll in corn syrup, then pecans.

## Pecan Candy

These candies are distinctive because they must be cut while still hot, and there is no way to avoid finger licking:

- 1/2 pound butter
- 2 cups flour
- 1 cup sugar
- 1 tablespoon cinnamon
- pinch salt
- 2 cups coarsely chopped pecans
- 1 egg, separated

Cream butter, sugar, and salt; blend in egg yolk; stir in flour and cinnamon. Spread mixture thinly on a jelly-roll sheet. Spread top with lightly beaten egg white. Cover with chopped pecans. Bake at 200–250°F for 45 minutes to one hour until firm and slightly browned on top. Remove and turn onto wax paper. Cut while hot. Store in a tin.

## Candied Orange Pecans

The southern recipe for Candied Orange Pecans fits into the domain of snack recipes. Many pecan snacks, whether sweet or savory, are claimed as regional specialties—often for several regions! Promotional recipe leaflets at conventions, chambers of commerce, or tourist offices usually recommend their own local pecans, and newspaper recipes share this bias.

- 12 ounces light brown sugar
- 8 ounces orange juice
- 8 ounces pecan halves
- grated rind of one large orange
- 1½ ounces butter

Grease a baking sheet. Stir sugar and juice in pan over a low heat until sugar is dissolved. Bring to a boil, and continue boiling until temperature reaches 236°F. As it thickens, test a few drops in a cup of cold water. When drops form small balls, take pan off heat, add grated rind and butter. Beat with wooden spoon until mixture begins to stiffen. Quickly add pecans and stir until they look slightly sugary. Turn onto the baking sheet and separate the nuts, using two forks or your fingers. Dry on a baking rack.

## Missouri Spicy-Sweet Pecans

- 1 cup sugar
- 1 teaspoon cinnamon
- 3/4 teaspoon salt
- 1 cup water
- 1 teaspoon vanilla
- 1 pound pecan halves

Toast pecans for 10 minutes at 300°F. Combine all ingredients except nuts and vanilla. Cook for about 5 minutes, until syrup spins a thread. Remove from heat. Add pecans and vanilla, and stir quickly until syrup crystallizes. Pour onto buttered plate. Separate nuts rapidly but gently.

## Georgia Roasted Treats

Toss 1 cup of unsalted Georgia pecan halves with 1 tablespoon vegetable oil mixed with 1 tablespoon Worcestershire sauce. Roast in a shallow baking pan at 275°F for 30 minutes, stirring every 5 minutes. Drain on paper towel and sprinkle lightly with salt.

## Guadalupe Valley Barbecued Pecans

- 2 tablespoons butter
- 1/4 cup Worcestershire
- 1 tablespoon ketchup
- 2 dashes hot sauce
- 4 cups pecan halves

Melt butter in Dutch oven, remove from heat and add seasoning. Add pecans and stir to coat. Bake in 15 x 10-inch pan at 400°F for 13 to 15 minutes, and shake gently every 5 minutes. Drain on paper towels and sprinkle with salt.

## Texas Spiced Pecans

- 1 egg white
- 1/2 teaspoon salt
- 1 tablespoon water
- 1 teaspoon cinnamon
- 3 cups pecan halves
- 1/4 teaspoon cloves
- 1/2 cup sugar
- 1/4 teaspoon nutmeg

Combine egg white and water. Stir in pecans and coat well. Combine remaining ingredients and sprinkle over pecans. Spread on well-buttered cookie sheet and bake at 300°F for 30 minutes, stirring two or three times.

## New Orleans Sugary Treat

- 1 pound pecan halves
- 1 egg white
- water
- 5/8 cup finest confectioner's sugar

Beat egg white with its weight in water till it forms a cream. With your fingers work in the sugar until it forms a smooth paste. Roll a small piece of this, place it between two pecan halves, then lightly roll in the paste, flattening the pecan somewhat. The coating outside must be very, very light, so that the delicate brown of the pecan meat shows through. Set the pecans to dry, and serve on dainty china saucers, setting a saucer to each guest.

▪ ▪ ▪

The second major category of pecan recipes involves pies. Many of these include a pastry shell with a morass of sticky, ultra-sweet syrup on which more or fewer pecans, depending on price and availability, are either randomly sprinkled or artistically patterned.

They are such staple fare that it is hard to imagine that their origins can be pinpointed. But there are claimants. One such is Slidell, Louisiana, once a small lumber town at the northeastern end of Lake Ponchartrain. It was named for John Slidell, a Confederacy ambassador, by his son-in-law, the railroad financier Baron Erlanger, who stopped there with his wife on the first train from Jackson, Mississippi, to New Orleans. Originally named Robert's Landing, it was the launching point for hunting trophies dispatched to New Orleans. A section of town running from the Pearl River to the center, Apple Pie Ridge, commemorates a woman who served apple pie to Civil War soldiers as they passed through the area on their way to battle. Whether pecans were added, or pecan pie was simply touted as part of a booster campaign addressed at likely settlers, remains a mystery. In the early years of this century, everyone knew this was where pecan pies originated. Nowadays, few have even heard the legend.

Vesta Harrison of Fort Worth, Texas, was another contender for the original pecan pie. As a teenager she attended a cooking school run by a certain Mrs. Chitwood from Chicago. One night, Vesta dreamed of a pecan pie. But Mrs. Chitwood brushed her off and told her there was no such thing. Undaunted, she said to herself, "Well, by gollies, I don't know how, but I'm gonna mess up something making a pecan pie." So she made a sorghum syrup pie and put a cup of pecans in. The teacher enjoyed it, asked for the recipe, and sent it off to an unrecorded organization in Washington. Very soon after, a five-hundred-dollar check was in the mail for her Texas Pecan Pie. The recipe, as adapted, follows:

## Texas Pecan Pie

- 1 scant cup sugar
- 1 heaping tablespoon flour
- 3 eggs
- 1 egg-sized chunk of butter
- 1 cup Karo syrup
- 1 teaspoon vanilla
- 1 cup pecans

Blend sugar and flour well. Pour mixture into a 9-inch pastry-lined pie plate and bake at 350°F for 45 to 50 minutes, or until filling is set.

■ ■ ■

When H. P. Stuckey, director of the Georgia Agricultural Experiment Station, wrote *Pecan-Growing* in 1925, he included a number of recipes that were "proposed, thoroughly tested, and found to be good by such authorities as Mrs. Thos. A. Banning, Elizabeth Wilson, Mrs. W. N. Hutt, Mrs. Harriet C. North, and Mrs. J. A. Kernoddle. They may be followed with assurance of good results." His pecan pie instructions were typical of no-frills family dishes:

## Pecan Pie (1)

Take 5 eggs, 1 cup molasses, 1 cup sugar, 2 tablespoons of flour, 1 cup pecans, 2 teaspoons butter. Beat the eggs light, add sugar, flour, molasses, butter and pecans. This makes two pies.

■ ■ ■

Pecan pies so frequently include Karo syrup that it is hard to believe the company's 1902 founding postdates the recipes by generations at the very least. This recipe, with or without the almond extract, is about as common as any:

## Pecan Pie (2)

| | |
|---|---|
| 1/2 cup sugar | 1/4 teaspoon salt |
| 2 tablespoons butter | 1 teaspoon almond extract |
| 2 eggs | 1 cup light Karo syrup |
| 2 tablespoons flour | 1½ cups chopped pecans |

Cream butter and sugar, add beaten eggs, flour, salt, extract, and syrup. Stir well. Add pecans, pour into crust, and bake half an hour in a moderate oven.

▪ ▪ ▪

Recipes in many cookbooks, such as *800 Proved Pecan Recipes*, which features 5,083 selected from 21,000 "favorite recipes," claim international origins, even German, Scandinavian, or Russian, although these countries are several thousand miles from the closest pecan trees. For example:

## Russian Tea Biscuits

| | |
|---|---|
| 2 sticks butter | 1/4 teaspoon salt |
| 2½ cups flour | 1 teaspoon vanilla |
| 1/2 cup powdered sugar | 1 cup pecan pieces |

Cream butter and sugar. Add other remaining ingredients. Roll into walnut-size balls and place on cookie sheet two inches apart. Flatten. Bake at 325°F for 15 minutes.

▪ ▪ ▪

The Texas Department of Agriculture, venturing into cuisine rather brusquely and short on details, also has a Russian recipe:

## PECANS

### Russian Rocks

2 cups pecans
1 package raisins
1 cup butter
1½ cups brown sugar
3½ cups flour
1 teaspoon cinnamon
1 teaspoon soda, dissolved in a little hot water

Make a stiff batter, drop by spoonfuls.*

■ ■ ■

Besides the pralines, candies, snacks, pies, and exotica, a host of recipes suggest pecans when almonds or walnuts would do, while in others the nuts merely add an extra crunch. There are innumerable recipes for salads, often congealed; bar cookies redolent with butter, sugar, and chocolate; and breads to use up seasonal gluts like cranberries.

Once in a while a recipe really stands out. It will be passed around on a three-by-five-inch file card, the back of an envelope, as a bridal-shower gift, at a dinner where one must impress, or to please someone loved. The following are two such recipes.

### Praline Cheesecake

1 cup graham cracker crumbs
3 eggs
3 tablespoons sugar
1½ teaspoons vanilla
3 tablespoons melted butter
1/2 cup finely chopped pecans
1½ cups dark brown sugar
maple syrup
2 tablespoons flour
whole pecans
3 eight-ounce packages cream cheese (room temperature)

Combine crumbs, sugar, and butter; press into 9-inch springform pan. Bake for 10 minutes at 350°F. Cream brown sugar, flour, and cream cheese, beating at medium speed till blended. Add eggs one

*The USDA unfortunately included no baking instructions, so some experimenting will probably be necessary using instructions from other cookie recipes.

at a time, mixing well. Blend in vanilla and chopped pecans. Pour over crumbs. Bake 50–55 minutes at 350°F. Loosen, cool before removing. Chill. Brush with maple syrup and garnish with pecan halves.

## Kahlua Cake

3/4 cup corn oil
1/3 cup Kahlua liqueur
4 eggs
1 teaspoon vanilla
1 cup sour cream
1 cup brown sugar
3/4 cup pecan pieces
One 3/4-ounce package vanilla instant pudding
1 package golden butter cake mix

Put cake mix in bowl; add and beat in eggs one at a time. Add sour cream, pudding mix, oil, vanilla. Divide in two. Pour one portion into greased 9-inch cake pan. To the remaining portion add brown sugar, nuts, liqueur. Mix well, then drop in heaping tablespoonfuls onto first portion, and swirl gently with knife to marble. Bake at 350°F for 75 minutes.

▪ ▪ ▪

Finally, and to get back to where it all began, an old Louisiana plantation cookbook offers a soup recipe that may well hark back to the Indian *powcohicoria*:

## Plantation Pecan Soup

2 cups finely ground pecans
2 cups water
3 stalks diced celery
3 tablespoons butter
3 tablespoons flour
salt and red pepper
2 cups chicken broth (if convenient)
3 cups sweet milk

Boil pecans in water. Cook, thickening with the butter and flour. Save a few pieces of pecan to sprinkle.

# ▪ CONCLUSION

Pecan trees are so cherished for the nuts they produce that it is easy to lose sight, metaphorically if not literally, of the trees. They are ideal for suburban and city landscaping, and their branches grace many suburban yards, providing leafy shade from summer sunshine. Generally, a single tree serves an average-size lot, set between fifty and eighty feet from the next tree, and thirty to forty feet from the tree trunk to the house itself in order to reach its full spread.

The trees are sturdy and resilient to ice or wind. In urban settings, their growth pattern allows the branches to be left low, to serve as a living wall, or pruned higher to make way for other flora. As the pecan has a long tap root, the tree does not compete with the root systems of other garden plants, and because the foliage appears later than it does on most trees, the leaves do not impede early flowering bulbs or the lawn grass. Where the sun is particularly savage, pecan trees will reduce the temperature in their shade by fifteen to twenty degrees. Less to their credit, although they are generous to compost heaps, the leaves can block gutters and make for laborious raking in the fall.

Pecan wood has numerous uses in manufacturing. It ranks third, behind black walnut and wild cherry, among the fine hardwoods. Useful for baseball bats, hammer handles, and agricultural implements, the wood is also chosen for furniture, wall paneling, flooring, and firewood. Occasional museum pieces and family heirlooms reveal its sturdy beauty. In Texas early woodworkers chose to work with the wood they found literally beside their doorsteps, and to fashion practical and beautiful objects still cherished generations later. One of these craftsmen, Albert Pawelek, was born in Poland in 1838 and came to Texas with his wife around 1860. A mail carrier and carpenter, Pawelek undertook woodworking as a hobby. Over the years he carved each of his eight children a polychromed pecan-and-pine crucifix. He carved altars for several area churches and a quantity of furniture. He lies buried in Panna Maria, the site of the first Polish settlement in the United States.

Although viewed by the pecan industry as mostly a nuisance, the shells also are processed to make furniture. Powdered into "flour," they serve as filler for plastic and veneer wood. Shells and defective "pops" are made into gravel for driveways, used as fuel, soil conditioner, stock bedding, and poultry litter, as filler for feeds, ingredients in insecticides, and fertilizers. They make a practical additive for adhesives, metal polish, non-skid paint, and dynamite. When fed through air intakes, the ground-up shells can speedily clean unwanted grease, bird feathers, and other potentially lethal debris from jet engines.

The tannic acid in the shells will acidify soils for plants like azaleas, magnolias, and camellias and provide mulch for ornamental plants. One early nurseryman in San Antonio even heard that a "decoction of the leaves is employed by the Mexicans as a hair-wash."

This last quirky use sums up the spectrum of pecan culture. On the one hand, the industry is an example of how high technology can overcome some of the procedures that caused difficulties early on, like shelling, storing, and transportation. On the other hand, the pattern of alternate bearing, in which one year's bountiful harvest is followed by a low yield, is nature's continuing tease. Pecan specialist George Ray McEachern, an extension horticulturist at Texas A&M University, sees "high-low swings in kernel availability" as a key problem needing to be resolved as soon as possible by a storage program, if not horticulturally.

The focus of the pecan industry in this century has moved not only technologically, but also geographically and culturally. Regional demand for pecans can be pinpointed to an exhibit of choice, thin-shelled pecans from the Mississippi bayous that was on display in New Orleans in 1886. When a more elaborate display of choice varieties was shown at the Paris Exposition in 1900, pecans attracted worldwide attention for the first time. Orders were placed by buyers from Greenland, Australia, Durban, South Africa Republic, and Ghorakhal, India. Twenty-five years later agricultural correspondents reported that most of the trees were still vigorous and productive.

In the United States, disease and infestation have pushed the industry westward. This trend has been encouraged by breeding programs that have produced cultivars adapted to a more arid environment.

An article in the *Country Gentleman* in 1914 bemoaned the fact that while northerners bought pecan orchards "somewhere down in Dixie, with trees all planted, under company care, developing paper wealth for the absentee owner," they rarely thought to buy the nuts. "People in New York City and in the North generally do not know good pecans. The kind they buy in the stores are looked upon in Texas as in the nature of a fraud, and a painted fraud at that, because they are thick-shelled mongrels . . . tinted and polished to give them surface attractiveness. It is almost impossible to extract their meat satisfactorily." The author, James H. Collins, was trying to promote the newly inaugurated parcel post as a means of selling pecans anywhere east of the Rockies at $2.50 to $3.00 for an eight- to ten-pound package—complete with nutcrackers.

Today, despite shelled nuts supplied to bakers and confectioners accounting for a high percentage of all pecan sales, customers who purchase only a pound or two of in-shell nuts are still respected. Alongside the highways from the south central to the southeastern states, roadside stands open their shutters each November. Drivers bring their cars to a halt and climb out, assuming they will be remembered from the previous year; often they are. One speaker at a Texas Pecan Growers' Association meeting, who had run a successful stand for years, admitted she kept a card file on each of her regular customers and kept track of their preferences. Her children grown, she was looking forward to her grandchildren helping out. Her advice, in addition to encouraging customers to sample varieties from the row of open bins, was simply, "Greet them with a hug!"

# RESOURCES

I found that researching the literature on pecans was not unlike approaching the tree itself, seeing the ripened clusters held closely together and yet still distinct from the next group. The reports of county agents, Ph.D. dissertations, travelers' tales, cookbooks, botanical treatises, conference proceedings, and an extraordinary assortment of learned and leisure reading formed separate clusters. Each had to be selected and considered in order to reap the full harvest.

One cluster was formed of books or chapters in books that provided sources of frequent reference: Jean Richardson Flack's "The Spread and Domestication of the Pecan (*Carya illinoensis*) in the United States" (Ph.D. dissertation, University of Wisconsin, 1970) gives an excellent geographical perspective. More than eighty years after publication, *Pecan-Growing* by H. P. Stuckey and E. J. Kyle (New York: Macmillan, 1925) continues to provide a sound framework. Each new edition of the *Texas Pecan Profitability Handbook* (College Station: Texas Agricultural Extension Service, Texas A&M University, 1990 edition edited by George Ray McEachern and Larry A. Stein; the 2007 edition by the same authors is retitled *The Texas Pecan Handbook*) is filled with articles by the cream of today's pecan experts. Tommy E. Thompson and L. J. Grauke detail genetic and horticultural traits in their contribution to *Acta Horticulturae*, "Genetic Resources of Temperate Fruit and Nut Crops: Pecans and Hickories (*Carya*)" (Wageningen, no. 290–XVIII, International Society for

Horticultural Science, 1991). J. Guy Woodroof's lengthy chapter on pecans in *Tree Nuts* (West, Conn.: Avi Publishing, 2d ed., 1979), has an especially good section about progress in storage techniques.

On several occasions early in this century, the pecan was featured in the *Yearbook of the United States Department of Agriculture* (Washington, D.C.: Government Printing Office, various dates). To single out a few of these articles, the "Promising New Fruits" round-up by William A. Taylor (1905, pp. 407–16) and several by unnamed authors (several years through 1912) were generally accompanied by one of E. J. Schutt's mouth-watering color plates; also interesting are Myer E. Jaffa's "Nuts and Their Uses as Food" (1906, pp. 295–312). and "Nut Breeding" by H. L. Crane et al. (1937, pp. 827–89). These, together with Rodney True's *Notes on the Early History of the Pecan in America* (Washington, D.C.: Smithsonian Report, 1917), cover the period that marked the industry's early years.

Selden C. Menefee and Orin C. Cassmore's *Pecan Shellers of San Antonio* (Washington, D.C.: USWPA Project, 1940) was very moving despite its official purpose and showed that *plus ça change, plus c'est la même chose*—the more things change, the more they stay the same.

Recipes formed another cluster and ranged from the Texas Department of Agriculture's slap-happy cookies and pies to the dishes served at the gatherings described by Linda West Eckhardt in *The Only Texas Cookbook* (Austin: Texas Monthly Press, 1986) and Terry S. Bertling in *Cooking with Pecans: Texas in a Nutshell* (Austin: Eakin, 1986).

For me, learning American history long after an English childhood, the readings were an opportunity to become better acquainted with a variety of people—like Thomas Jefferson, whom I encountered on the pages of his garden book (Edwin Morris Betts's annotated *Thomas Jefferson's Garden Book* [Phila-

Nutcrackers—such as this distinctive model—are essential tools for separating the delicious pecan kernel from its shell. *Photo courtesy USDA.*

delphia: American Philosophical Society, 1944]) and many American Indians in Grant Foreman's vivid account *The Five Civilized Tribes* (Norman: University of Oklahoma Press, 1934). I traveled the frontier vicariously through Reuben G. Thwaites's thirty-two volumes of *Early Western Travels: 1748–1846* (Cleveland, Ohio: Arthur H. Clark, 1904–1907) and was tempted by authors like Stanley Arthur, who described *Old New Orleans: A History of the Vieux Carré* (New Orleans: Harmandson, 1944) and John W.

Monette and his *History of the Discovery and Settlement of the Valley of the Mississippi* (New York: Harper and Brothers, 1846).

In writings about Texas pecans, sequential versions of J. H. Burkett's Texas Department of Agriculture bulletins (Austin: Texas Department of Agriculture, Bulletin no. 77 and others) were invariably enjoyable, especially the seemingly off-the-cuff comments that he included from his county agents on the new cultivars being tested in their areas. The proceedings from early Texas Pecan Growers' Association conferences were always a treat. No one writes comparable papers today, sprinkling sly wit through a mixture of homily, confessional, and scientific update. Many of the participants were legends long before they nourished the roots of the pecans they raised, among them E. E. Risien (who wrote for many decades), J. H. Burkett, H. P. Attwater, both F. M. Ramsey (Lampasas) and F. T. Ramsey (Austin), H. A. Halbert, W. J. Millican, and E. W. Kirkpatrick.

However useful and appropriate a reference turned out to be, nothing quite matched the excitement of finding a book no one else seemed to know about. I found Willie Mac Weinert's *Authentic History of Guadalupe County* (Seguin, Tex.: Enterprise Publishers, 1951) one fall afternoon in the Seguin public library. I had ridden over to see the pecans, mile after mile of trees ready for harvesting, but became captivated by the small book. It was everything a county history should be, with slightly blurred black-and-white photographs and wondrous tales of early characters who came from far away to settle beside the Guadalupe River. Many names are now forgotten, but their legacy remains in the county's prolific orchards and groves.